FRACTIONS

AND MORE

How to make your child successful in Math

Dina Migachyov

Quaternion Press

Cover and illustrations by Larisa Migachyov

Published by Quaternion Press Publishing House
P.O. Box 700564
San Antonio, Texas 78270, USA

ISBN 0-967-25351-9
Printed in the United States of America

FRACTIONS AND MORE

INTRODUCTION

"Fractions and more" is the second book of a two book set. The first book, "Mathematics for Little Ones" covers addition, subtraction, multiplication, and division of whole numbers. After mastering this material it is time to start studying fractions.
Fractions are very important part of our every day life and extremely important part of math. It is absolutely impossible to study high level math without knowing fractions very well.
In the beginning "Fractions and more" teaches children to work with big numbers, then gives them theory of numbers, and then, only after they are prepared for this, teaches them fractions. I think that we should not teach decimals before fractions or separately from them. Decimals are specific instance of fractions, and this book presents them as specific case of fractions.
"Fractions and more" teaches all four arithmetic operations with fractions, decimals, and mixed numbers. Besides, it teaches how to combine all arithmetic operations with all kind of numbers. I don't know why we stopped doing it in America. I think, that it is very important topic, and children in all civilized countries study it very seriously.
Like the first book, "Fractions and more" gives a lot of word problems, which become more complicated and sophisticated. Solving them prepares children for solving real life problems, which definitely requires using all kind of numbers.
Parents can start using this book after their child mastered four arithmetic operations with whole numbers. It means, that, in general, this book is for children from 4th to 6th grade; but if your child can start it earlier, it will be good. But if you see that your son or daughter is older, maybe even in high school, and can not operate freely with fractions, decimals, or mixed numbers, encourage him or her to repeat or learn this material using this book.

Work hard and have fun!

How to Use This Book

The first part of this book is a textbook, and the second part is a workbook. The workbook contains exercises for each lesson and a chapter test for each chapter. It is absolutely necessary to do all the exercises after each lesson – do not go to the next lesson until your child has mastered the previous lesson's material.

Your child should be ready for this book after finishing "Mathematics for Little Ones". If he or she is more advanced, find out where to start by giving him or her each chapter test and seeing where he or she has difficulty. Start with the first chapter where your child has less than 100% on the chapter test – even if the score is 19 out of 20. Then, start using this book, doing all the exercises and tests at your child's own pace regardless of what material is being covered in school. If the school is good, you will find that some of the material will be covered in class. In that case, just do the chapter test and move on. If you are not that lucky and your child is doing some other material in school, don't worry. Just study this book with your child. No matter what your child is studying in school, when you finish this book, your child will master fractions and be fully prepared to go on to pre-algebra and more advanced mathematics.

If your child is in a homeschooling environment, this book is all you need to make sure your child masters all the necessary arithmetical skills. Just go with your child chapter by chapter, doing all the exercises and tests at his or her own pace. In your home you don't have to worry about the math program your school district has selected, or about the latest educational fads. This book covers the basic material your child will need for the math part of the SAT test. You should also supplement this book with books on the history of mathematics and books of fun mathematical puzzles (such as the many wonderful books of Martin Gardner). Check our website at www.quaternionpress.com to find some book recommendations.

TABLE OF CONTENTS

WORKBOOK

Chapter 1
Multiplication and Division of Large Numbers

Lesson 1
Multiplication of Numbers Ending in Zeros

You already know the table of multiplication. If you can't quite remember it, don't worry; simply review Chapter 4 in <u>Mathematics for Little Ones</u>.

Now, we will learn how to multiply any two whole numbers. We will start with easy problems and go on to harder ones.

Let us start with the multiplication of a 2-digit number by a 1-digit number, such as 32 x 2. To solve such an easy problem, you can use simple logic: 32 has 3 tens and 2 units; if we take it twice, we will have 6 tens and 4 units, so the answer is 64. Unfortunately, it is not so easy every time. For example, if you have to multiply 76 x 9, you have to do a more complicated calculation. In this chapter, I will show you how do this calculation easily.

Let us begin with the simplest case – how to multiply a number ending in zeros by a one-digit number. You already know that when we multiply any number by 10, the result is just that number with a 0 on the end; 5 x 10 = 50. Suppose we are multiplying 5 x 30. Think about it like this: you take 3 tens 5 times. The result should be 15 tens, or 150. In other words, we can multiply the nonzero digits and write 0 at the end.

There are two ways to write it: 1) 5 x 30 = 150 and 2)

$$\begin{array}{r} 5 \\ \underline{\times\ 30} \\ 150 \end{array}$$

This will also work if you have more than one zero at the end of a number; 6 x 1,000 = 6,000. After all, all you are doing here is taking a thousand 6 times; of course, you get six thousand. So,

it is easy to see that if you multiply a number ending in many zeros by a one-digit number, you get the product of the two nonzero digits with the same number of zeros at the end.
For example, let us multiply 7 by 300.

$$\begin{array}{r} 7 \\ \times\ \underline{300} \\ 2100 \end{array}$$

Look at these examples, and you will see that it is very easy.

3 x 300 = 900 [3 times 3 hundreds gives us 9 hundreds]
80 x 4 = 320 [4 times 8 tens gives us 32 tens]
2,000 x 7 = 14,000 [7 times 2 thousands gives us 14 thousands]

This is easy; but what if both numbers end in zeros? This is not hard, either. Think of what happens when we multiply, for example, 10 by 100. 10x100=1000; ten hundreds is the same thing as one thousand. You see that the product of these numbers has as many zeros at the end as both of the numbers we're multiplying. So, if we have two numbers that each end in zeros, all we do in this case is multiply the nonzero digits, and write all the zeros from both numbers at the end. Here is how to do this:

$$\begin{array}{r} 300 \\ \times\ \underline{40} \\ 12000 \end{array} \qquad \begin{array}{r} 80 \\ \times\ \underline{500} \\ 40000 \end{array} \qquad \begin{array}{r} 600 \\ \times\ \underline{400} \\ 240000 \end{array}$$

Now, practice.

PRACTICE

$$\begin{array}{r} 200 \\ \times\ \underline{300} \end{array} \qquad \begin{array}{r} 6000 \\ \times\ \underline{400} \end{array} \qquad \begin{array}{r} 70 \\ \times\ \underline{5000} \end{array} \qquad \begin{array}{r} 500 \\ \times\ \underline{90} \end{array} \qquad \begin{array}{r} 8 \\ \times\ \underline{800} \end{array} \qquad \begin{array}{r} 900 \\ \times\ \underline{80000} \end{array}$$

Lesson 2
Multiplication of 2-digit Numbers by 1-digit Numbers

Now, let us try something harder. What if the two digit number has no zeros? That is, suppose we have to do a problem like this: 76 x 9. This looks hard until we remember how to write numbers in expanded form. If we write 76 as 70 + 6 , we can think of our problem as taking 70 nine times, taking 6 nine times, and then adding the two answers. You already know how to do 70 x 9; the answer to that is 630. 6 x 9 = 54. So, our answer is the sum of 630 and 54, or 684. Here is how we write it:

At first, write the numbers one under another, lining them up by place value:

$$\begin{array}{r} 76 \\ \text{x} \quad 9 \end{array}$$

Then, we multiply the units digit by 9; 6 x 9 = 54.
We write the 4 under the units digits and write the 5 above the tens digits, like this:

$$\begin{array}{r} 5 \\ 76 \\ \text{x} \quad \underline{9} \\ 4 \end{array}$$

Now, we multiply the tens digit by 9; 7 x 9 = 63. We also add the 5 tens that we got as a result of our first multiplication; 63 + 5 = 68. Now, we write the 68 in the tens position of our answer.

$$\begin{array}{r} 5 \\ 76 \\ \text{x} \quad \underline{9} \\ 684 \end{array}$$

Let us try another example:

$$\begin{array}{r} 37 \\ 39 \\ \text{x} \quad \underline{8} \\ 312 \end{array}$$

Before we finish the lesson, let us write the algorithm for the multiplication of a 2-digit number by a 1-digit number:

1. Multiply the number of ones by the one digit number and write the number of ones in the result under the line in the ones position. Carry the number of tens (if there is one) to the tens position and write it above the tens position of the top number.

2. Multiply the number of tens by the one digit number, then add the result of that multiplication and the number which was carried from the previous position (if there is one), and write the result under the line.

Now look at these examples, do the problems yourself and compare your answers with the ones in the book. If all the answers are correct, try the practice problems. If not – start this lesson over.

	2	3	2	4		4
34	28	39	45	76	81	96
x 2	x 3	x 4	x 5	x 8	x 9	x 7
68	84	156	225	608	729	672

Now practice.

PRACTICE

41 x 2	23 x 3	25 x 3	37 x 4	64 x 5	58 x 6	47 x 2
72 x 7	13 x 5	29 x 4	84 x 2	91 x 9	35 x 8	78 x 5
32 x 2	34 x 3	61 x 4	11 x 7	23 x 5	52 x 6	74 x 3

Lesson 3
Multiplication of Large Numbers

If you understand how to multiply 2-digit numbers and have had enough practice, this lesson will be easy and fun for you. The algorithm of multiplication is absolutely the same: the number of units should stay in the same position, but the number of tens should be sent to the next one.

Let us start with the multiplication of a 3-digit number by a 1-digit one.

```
1 3 2
  2 5 4
x     7
1,7 7 8
```

4 x 7 = 28; so, we write 8 in the ones position and send 2 to the tens position. We will write this number above the tens position, so we don't forget to add it later.
5 x 7 = 35; we have 35 tens. We will not forget to add the 2 tens that we had carried from the previous position, and get 37 tens, which means 3 hundreds and 7 tens. Now we will write the 7 under the line in the tens position, and send 3 to the hundreds position.

2 x 7 = 14; we have 14 hundreds, which means 1 thousand and 4 hundreds. We will add 14 and 3 to find the number of hundreds, and send the 1 to the thousands position.

Look at these examples and try to do them yourself. Then compare the results.

$$\begin{array}{r} \\ 231 \\ \text{x}\quad 3 \\ \hline 693 \end{array} \qquad \begin{array}{r} {\scriptstyle 4\ 3} \\ 705 \\ \text{x}\quad 7 \\ \hline 4935 \end{array} \qquad \begin{array}{r} {\scriptstyle 212} \\ 627 \\ \text{x}\quad 4 \\ \hline 2508 \end{array} \qquad \begin{array}{r} {\scriptstyle 243} \\ 586 \\ \text{x}\quad 5 \\ \hline 2930 \end{array} \qquad \begin{array}{r} {\scriptstyle 83} \\ 894 \\ \text{x}\quad 9 \\ \hline 8046 \end{array}$$

If you got the right answers, you are ready to learn to multiply larger numbers. It will take longer to do the problems, but the idea is the same: the number of ones should stay in the same position, the number of tens should be sent to the next one. So that you don't forget to add them, you should write the number above the next position.

Look at these examples, do them yourself and compare the results.

$$\begin{array}{r} \\ 3,424 \\ \text{x}\qquad 2 \\ \hline 6,848 \end{array} \qquad \begin{array}{r} {\scriptstyle 2\ \ 2\ \ 1\ 1} \\ 7,856 \\ \text{x}\qquad 3 \\ \hline 23,568 \end{array} \qquad \begin{array}{r} {\scriptstyle 1\ \ 2\ 4\ 5\ 5} \\ 12,567 \\ \text{x}\qquad 8 \\ \hline 100,336 \end{array} \qquad \begin{array}{r} {\scriptstyle 1\ 4\ 2\ 5\quad 6} \\ 263,809 \\ \text{x}\qquad 7 \\ \hline 1,846,663 \end{array}$$

You see that you can multiply very large numbers and it is not so hard. The next example shows that you can even multiply millions.

Look at the example:

$$\begin{array}{r} {\scriptstyle 1\ 1\ 1\quad 1\ \ 2\ 2} \\ 23,542,367 \\ \text{x}\qquad\qquad 3 \\ \hline 70,627,101 \end{array}$$

Here is what you should do:

3 x 7 = 2 1; 1 will go in the ones position, 2 will be sent to the tens position;
6 x 3 = 1 8, 18 + 2 = 20; 0 will go in the tens position, 2 will be sent to the hundreds position;
3 x 3 = 9, 9 + 2 = 11; 1 will go in the hundreds position, 1 will be sent to the thousands position;
2 x 3 = 6, 6 + 1 = 7; 7 will go in the thousands position, nothing will be sent to the next position;
4 x 3 = 12; 2 will go in the ten-thousands position, 1 will be sent to the hundred-thousands position;
5 x 3 = 15, 15 + 1 = 6; 6 will go in the hundred-thousands position, 1 will be sent to the million position;
3 x 3 = 9, 9 + 1 = 10; 0 will go in the millions position, 1 will be sent to the ten-million position;

2 x 3 = 6, 6 + 1 =7; 7 will go in the ten-millions position, nothing will be sent to the next position.

As you can see, it's time-consuming, but not very hard, and it definitely can be fun!
Now practice, but start your practice with smaller numbers.

PRACTICE

$\begin{array}{r}232\\ \underline{\times\ 2}\end{array}$ $\begin{array}{r}134\\ \underline{\times\ 3}\end{array}$ $\begin{array}{r}261\\ \underline{\times\ 4}\end{array}$ $\begin{array}{r}311\\ \underline{\times\ 7}\end{array}$ $\begin{array}{r}423\\ \underline{\times\ 5}\end{array}$ $\begin{array}{r}152\\ \underline{\times\ 6}\end{array}$ $\begin{array}{r}274\\ \underline{\times\ 3}\end{array}$

$\begin{array}{r}372\\ \underline{\times\ 5}\end{array}$ $\begin{array}{r}2{,}453\\ \underline{\times\ \ \ 2}\end{array}$ $\begin{array}{r}5{,}725\\ \underline{\times\ \ \ 7}\end{array}$ $\begin{array}{r}32{,}812\\ \underline{\times\ \ \ 6}\end{array}$ $\begin{array}{r}63{,}293\\ \underline{\times\ \ \ 8}\end{array}$ $\begin{array}{r}17{,}604\\ \underline{\times\ \ \ 9}\end{array}$

$\begin{array}{r}352{,}467\\ \underline{\times\ \ \ 4}\end{array}$ $\begin{array}{r}478{,}832\\ \underline{\times\ \ \ 8}\end{array}$ $\begin{array}{r}23{,}456{,}271\\ \underline{\times\ \ \ 9}\end{array}$ $\begin{array}{r}45{,}258{,}107\\ \underline{\times\ \ \ 7}\end{array}$

Lesson 4
Division of 2-digit Numbers by 1-digit Numbers

To be able to divide large numbers, you should remember all the division facts (if you forgot them, review Chapter 5 of *Mathematics for Little Ones*) and understand place values of whole numbers (Lesson 1, Chapter 6, *Mathematics for Little Ones*).
Suppose you want to divide 48 by 2. 48 is the dividend, 2 is the divisor, and the result of the division is the quotient. Let us break the dividend into its expanded form:

$$48 = 40 + 8$$

Then, you can divide 40 by 2 and divide 8 by 2 and add the two answers to get the result.

40 is 4 tens; if we divide 4 tens by 2, the result will be 2 tens.

$$40 \div 2 = 20$$

You already know that

$$8 \div 2 = 4$$

Now, we will add the number of twos in 40 and the number of twos in 8.

$$20 + 4 = 24$$

We can write: $48 \div 2 = 24$

That was not so hard, but unfortunately, we are not so lucky every time. We often can't divide each position evenly. For example, in calculating $51 \div 3$, we can't divide 50 by 3 without a remainder. That's why we usually use a different method for division.

Do you remember how to divide with a remainder? If you forgot, review the material from Lesson 11, Chapter 5, *Mathematics for Little Ones*.
Now I want to show you 3 ways of writing division. For example, if we want to divide 6 by 3, we can write it as we have been doing all this time: $6 \div 3$, or we can write it like this: $\frac{6}{3}$, or

we can write it like this: $3\overline{)6}$.

The third way is the most convenient for dividing large numbers. But be careful – the divisor is written before the dividend; think of it as trying to find out how many times 3 goes into 6.
Let us divide 78 by 3. The result of the division will be on top.

$$\begin{array}{r}
2 \\
3\overline{)78} \\
\underline{6} \\
1
\end{array}$$

At first, let us divide 7 tens by 3. We will get 2. We multiply the 2 by 3 and get 6, which we write under the 7. Subtracting 6 from 7, we find that the remainder is 1. It means that we have 2 tens in the quotient. We write it above the 7 and continue the division.

Now, we have 1 ten left as a remainder and 8 units that we have not divided yet. Let us put them together; we will get 18. (we will just write the 8 next to the 1). All we have to do now is divide 18 by 3. The answer is 6, which we write next to the 2 in the quotient. Multiplying 3 by 6, we get 18, which we write under the 18 units in our problem. When we subtract 18 from 18, we get 0, which means that we have divided evenly, without a remainder. Since we don't have any more digits left, we are done; the answer is 26.

$$\begin{array}{r}
26 \\
3\overline{)78} \\
\underline{6} \\
18 \\
\underline{18} \\
0
\end{array}$$

Before you start practicing, I will give you the algorithm for dividing a 2-digit number by a 1-digit one. I will give you this algorithm with an example to illustrate division.

```
     1 5
  5 ) 7 5
      5
      2 5
      2 5
        0
```

1. Divide the number of tens by the divisor (7 ÷ 5); write the quotient above the tens.
2. Multiply the quotient by the divisor and subtract the result from the number of tens (1 x 5 = 5; 7 - 5 = 2)
3. Compare the remainder with the divisor; it should be less than the divisor. (2<5)
4. Bring down the number of ones, write it next to the remainder (you get 25).
5. Divide this new number by the divisor; write the quotient (5) above the ones
6. Multiply the quotient (5) by the divisor and subtract the result from the number of ones. (5 x 5 = 25; 25 – 25 = 0)
7. Compare the remainder with the divisor; it should be less than the divisor.

If the remainder equals zero, we can say that the number is divisible by the divisor. If it does not equal zero, we write the remainder next to the result, as we have been doing before.

Look at these 2 examples, one without a remainder and one with a remainder.

```
      24              12 R 1
  4 ) 96          7 ) 85
      8               7
      16              15
      16              14
       0               1
```

PRACTICE

LEVEL 1

2) 48 3) 63 4) 88 2) 64 3) 96 4) 40

2) 64 3) 39 4) 44 2) 80 3) 63 4) 84

LEVEL 2

3) 54	4) 64	5) 85	2) 78	6) 72	7) 84
8) 96	3) 84	4) 76	5) 60	7) 91	6) 96

Lesson 5
Division of Large Numbers by 1-digit numbers

Now that you know the algorithm for dividing 2-digit numbers by 1-digit numbers, it will not be difficult to divide 3-digit, 4-digit and even longer numbers by 1-digit numbers. The main thing is that the algorithm of division is the same:

1) divide
2) multiply
3) subtract
4) compare
5) bring down the next position
6) return to step 1.

Let us divide a 3-digit number using this algorithm: 561 ÷ 3.

```
     1 8 7
3 ) 5 6 1
   -3
    2 6
   -2 4
      2 1
     -2 1
        0
```

1) 5 ÷ 3 = 1 (R 2, we will write 1 on the top above 5)
2) 1 x 3 = 3
3) 5 - 3 = 2
4) 2 < 3
5) bring down 6 and we will have 26 tens

1) 26 ÷ 3 = 8 (R 2,we will write 8 on the top above 6)
2) 8 x 3 = 24
3) 26 - 24 = 2
4) 2 < 3
5) bring down 1 and we will have 21 tens

1) 21 ÷ 3 = 7 (we will write 7 on the top above 1)
2) 7 x 3 = 21
3) 21 - 21 = 0
4) 0 < 3
5) we don't have anything to bring down – we are done with the division.

You can see that it is not harder than dividing 2-digit number, but it is a little bit more time-consuming. Now, let us try a 5-digit number; let us divide 13,756 by 4.
1 is less than 4, so we can start the division with 13 thousand.

```
       3, 4 3 9
4 ) 1 3, 7 5 6
   - 1 2
       1 7
     - 1 6
         1 5
       - 1 2
           3 6
         - 3 6
             0
```

1) 13 ÷ 4 = 3 (R 1, write 3 above 3)
2) 3 x 4 = 12
3) 13 - 12 = 1
4) 1 < 4
5) bring down 7; we have 17 hundreds

1) 17 ÷ 4 = 4 (R 1, write 4 above 7)
2) 4 x 4 = 16
3) 17 - 16 = 1
4) 1 < 4
5) bring down 5; we have 15 tens

1) 15 ÷ 4 = 3 (R 3, write 3 above 5)
2) 3 x 4 = 12
3) 15 - 12 = 3
4) 3 < 4
5) bring down 6; we have 36 ones

1) 36 ÷ 4 = 9 (write 9 above 6)
2) 9 x 4 = 36
3) 36 - 36 = 0
4) 0 < 4
5) we don't have anything to bring down; we are done.

Before you start practicing, I want to show you a special case that can be confusing – the case when, after bringing down a number, you still get a number smaller than the divisor.
Look at this example: 545 ÷ 5.

```
     1 0 9
5 ) 5 4 5
   - 5
      4
    - 0
      4 5
    - 4 5
        0
```

You can see that after dividing 5 hundred by 5, we got 1 and didn't get any remainder (you don't have to write the 0 when you subtract), but the number of tens is less than 5. We can't divide 4 into 5 equal parts. In this case we write 0 in this position, and then continue the division as we did before.

Now look at these examples and then do the problems yourself.

```
    2 2 9          2 5 9            7 7 3             3 5, 0 1 7
2 ) 4 5 8      3 ) 7 7 7      5 ) 3, 8 6 5      7 ) 2 4 5, 1 1 9
    4              6               3 5              2 1
      5            1 7               3 6              3 5
      4            1 5               3 5              3 5
    1 8              2 7               1 5                1
    1 8              2 7               1 5                0
      0                0                 0                1 1
                                                            7
                                                            4 9
                                                            4 9
                                                              0
```

Now practice.

PRACTICE

2) 3 5 8   3) 2, 6 3 1   4) 5, 1 7 2   5) 2 6, 1 9 5

3) 5 2, 1 0 4   7) 1 8, 5 6 4   6) 7 7, 2 0 2   8) 1 2 3, 0 24

2) 3 4 5, 6 7 8, 9 1 0   3) 1 3 5, 7 9 0, 2 4 2   5) 8 2 5, 4 3 1, 2 4 5

Lesson 6
Multiplication by 2-digit Numbers

You know the idea of multiplication: if we want to add the same number several times, we use multiplication instead. We don't add 7 + 7 + 7 + 7 + 7 + 7 + 7 + 7; we multiply 7x 8.
If we have to multiply 35 by 23, it means that we have to add it 23 times. We can do it in two steps; we can add it 20 times, and then 3 more times. So, we can multiply 35 by 20, multiply 35 by 3, and then simply add the results.

You already know from Lesson 1 how to multiply 35 by 20.

35 x 20 = 700 35 x 3 = 105 700 + 105 = 805

To save time, we will write it like this:

$$\begin{array}{r} 35 \\ \times\ \underline{23} \end{array}$$

At first, multiply 35 by 3 and write the result under the line, like this:

$$\begin{array}{r} {}^{1} \\ 35 \\ \times\ \underline{23} \\ 105 \end{array}$$

Now multiply 35 by 20: to save work, multiply 35 by 2 and write a 0 at the end.
For your convenience, erase the 1 from the previous multiplication; think only about the tens:

$$\begin{array}{r} 35 \\ \times\ \underline{23} \\ 105 \\ \underline{700} \end{array}$$

Now, add the results of the multiplication.

$$\begin{array}{r} 35 \\ \times\ \underline{23} \\ 105 \\ +\ \underline{700} \\ 805 \end{array}$$

Actually, I recommend you to write a 0 in the ones position at once when multiplying by the tens digit – you know that there will be a 0 there every time.

Now, let us do another example. For now, I will write it in steps.

	1)	2)	3)
	1 2 3	1 3 4	
2 5 8	2 5 8	2 5 8	2 5 8
x 6 4	x 6 4	x 6 4	x 6 4
	1 0 3 2	1 0 3 2	1 0 3 2
		1 5 4 8 0	+ 1 5 4 8 0
			1 6 5 1 2

To make it easier, erase or cross out the numbers on top after multiplying by the ones before you start multiplying by the tens.

2 3	3 4	5 5	1 1 1
~~1~~	~~4 4~~	~~2 2~~	~~2 3 3~~
47	56	89	678
x 52	x 78	x 63	x 24
94	448	267	2712
+ 2350	+ 3920	+ 5340	+ 13560
2444	4368	5607	16,272

If you have to multiply a number that ends in a zero by a 2-digit number, like 30 x 47, use the commutative property of multiplication: change the problem in 47 x 30. In this case, it is much easier to find the product. Multiply 47 by 3 and then write a zero on the right of the product.

2
47
x 30
1610

Now practice.

PRACTICE

42	56	84	67	29	38
x 30	x 70	x 60	x 50	x 40	x 90

5 8	7 9	8 5	4 3	6 2	9 4
x 2 7	x 3 6	x 7 3	x 4 8	x 7 5	x 2 2

2 3 8	6 4 2	5 8 7	2 8 9	7 2 3	4 0 4
x 2 6	x 5 6	x 7 2	x 4 9	x 8 6	x 9 3

Lesson 7

Multiplication by 3-digit Numbers

If you understand how to multiply by 2-digit numbers, this lesson will be easy and fun. Recall that we can write any 3-digit number in expanded form.

$$325 = 300 + 20 + 5.$$

You already know how to multiply by 25; the only new thing to learn is how to multiply by 300. I think that you can understand that instead of multiplying by 300, you should multiply by 3 and then write two zeros at the end of the result. Similarly, for any other number of hundreds, multiply by the number of hundreds and write two zeros at the end of the result.

Now, let us multiply 135 by 247. We will write the numbers vertically and go step by step:

```
              23         12          1
  135        135        135        135        135
x 247      x 247      x 247      x 247      x 247
             945        945        945        945
                       5400       5400     + 5400
                                 27000      27000
                                            33345
```

In real life you will not have to write the problem 5 times, as I do here. Your multiplication will look like the final step. Remember to start multiplying by any new position by writing one or two zeros and erasing all the numbers above the first factor.
Now look at these two examples and then start practicing.

```
      1 1           1   4
      2 1           1   5
      3 2               1
      1 8 6         1,2 0 9
    x 2 3 4       x   5 6 2
      7 4 4         2 4 1 8
    5 5 8 0       7 2 5 4 0
 + 3 7 2 0 0   + 6 0 4 5 0 0
   4 3,5 2 4     6 7 9,4 5 8
```

If you have to multiply a number with fewer digits by one with more digits, use the commutative property of multiplication and change the order. For example, don't multiply 23 by 567 – multiply 567 by 23. It will save you one step and the problem will be easier.

PRACTICE

24 x 300	237 x 700	650 x 500	1,287 x 600
39 x126	248 x 534	812 x 907	2,508 x 246
146 x248	672 x 159	701 x 627	4,819 x 305

Lesson 8
Division by 2-digit numbers

It makes no difference whether you divide by a 1-digit or 2-digit number; the algorithm of division is absolutely the same. However, division by a 2-digit number is much more time-consuming and requires much more multiplication. Let us work with some examples.
For example, suppose you want to divide 888 by 37.
Write the divisor outside the bracket and the dividend inside:

$$37\,\overline{)\,888}$$

Then, start the division. 8 is smaller than 37, so we will start with 88 tens. Now, what we have to do is figure out how many times 37 can go into 88. Let us multiply it by some numbers to see how close we can get.

37 x 1 = 37
37 x 2 = 74
37 x 3 = 111

You can see that 2 is the only good quotient because 74 is closer to 88 than 37, but 111 is already too big.
Now, multiply 2 by 37 (2 x 37 = 74), subtract the result from 88 (88 - 74 = 14, 14 < 37) , and bring down the next digit, 8.

```
      2
37 )888
  - 74
    148
```

Now you have to figure out how many times 37 goes into 148. You already know that 37 x 3 = 111, which is too small. Let us try something bigger.

```
  2
  37
x  4
 148
```

37 x 4 = 148. It means that 37 goes into 148 exactly 4 times, i.e. we can write 4 above the line in the ones position. Then, again, we multiply, subtract, compare and we are done.

```
     24
37 ) 888
   - 74
     148
   - 148
       0
```

We are not so lucky every time; sometimes, we have a remainder. For example:

```
       12 R5
54 ) 653
    -54
     113
    -108
       5
```

You can see that 54 goes into 65 only once, so we have 1 in the quotient in the tens position. Then we do the following:

1) 54 x 1 = 54
2) 65 - 54 = 11 (11 < 54)
3) write down 3 and get 113 ones
4) 54 x 1 = 54, 54 x 2 = 108, 54 x 3 = 162 >113

108 is the closest to 113, and so we
will write 2 in the quotient in the ones position

1) 54 x 2 = 108
2) 113 - 108 = 5 (5 < 54)
3) we don't have anything else to bring down; we are done!

5 is the remainder. We write it near the whole part of the quotient.

The most difficult and time-consuming thing is finding how many times a divisor goes into a dividend or a part of a dividend. In the previous example, it was not hard to see that 54 goes into 65 only once. But sometimes, it is not so obvious. For example, let us divide 744 by 12.

```
    ____
12 ) 744
```

How many times does 12 goes into 74? Don't start by multiplying by 1, by 2 and so on. It will take too much time. Instead, compare the numbers of tens: 7 and 1.

Since 7 is 7 times bigger than 1, start with 7.

12 x 7 = 84 > 74

Because it is too big, try a smaller number such as 6.

12 x 6 = 72 < 74

72 is smaller than 74 and the difference between them is smaller than 12 (74 - 72 = 2). It means that 6 is the correct number.

```
      62
12 ) 744
    -72
      24
     -24
       0
```

It is not hard to see that 12 goes into 24 exactly 2 times.

This topic is not very easy. I will give you several examples, but I want you to solve them yourself and only compare the results when you are done.

```
       2 3             3 3               7 3           2 0 1 R 12
1 8 ) 4 1 4    2 3 ) 7 5 9    1 4 ) 1 0 2 2    4 5 ) 9,0 5 7
     - 3 6          - 6 9          -  9 8            - 9 0
       5 4            6 9             4 2                5
     - 5 4          - 6 9           - 4 2              - 0
         0              0               0                5 7
                                                       - 4 5
                                                         1 2
```

Now practice.

PRACTICE

$12\overline{)288}$	$18\overline{)648}$	$23\overline{)943}$	$35\overline{)7{,}210}$	$41\overline{)5{,}043}$
$24\overline{)1{,}248}$	$31\overline{)1{,}209}$	$52\overline{)728}$	$92\overline{)1{,}012}$	$67\overline{)2{,}345}$
$18\overline{)432}$	$24\overline{)792}$	$47\overline{)5{,}687}$	$56\overline{)11{,}984}$	$77\overline{)4{,}312}$

Lesson 9
Averages

Suppose your class took a math test and everyone did very well; some people got 80 points, some people got 90 points, and some people got 100. Another class in your school, however, was not so lucky; they got a very hard test, and most people got 60, 70, or 80 points. I think you can understand that a score of 80 points is not a very good score in the first class, while that same score of 80 is very good if you are in the second class. How can we define this mathematically? We use a number called the **average.** This number shows the score that is neither very good nor very bad, but somewhere in between. Then we can say that if you got a score that is above average, you did well; and if you got a score below average, you did badly. If the average is 90, a score of 80 is terrible; if the average is 70, that same score of 80 is terrific.

But how do we calculate the average? Fortunately, there is an easy way of doing it. Suppose you have 4 numbers: 3, 9, 2, 6. To find the average of these numbers, all you do is add the numbers together and divide the result by how many numbers you have. In our example, the average will be $(3 + 9 + 2 + 6) \div 4 = 20 \div 4 = 5$.

Let us find another average. This time, our numbers will be: 5, 13, 17, 24, and 36. The average is $(5 + 13 + 17 + 24 + 36) \div 5 = 95 \div 5 = 19$.

I think you understand that the average can't be greater than the greatest number or smaller than the smallest one; it should be somewhere in between. Sometimes the average is not a whole number; but since you have not studied fractions yet, I will not give you this kind of numbers. Let us try some problems.

PRACTICE

Find the average of these numbers.

1. 25, 37
2. 8, 7, 12, 5
3. 14, 24, 34, 44
4. 28, 19, 31
5. 11, 12, 13, 14, 15
6. 23, 49

Lesson 10
Word problems

1.If one page of a book has approximately 40 lines and each line has approximately 56 characters, how many characters does the page have?
2. If one page of a book has approximately 2240 characters, and the book has 72 pages, how many characters does the book have?

3. There are 35 chocolate-covered peanuts in one pack. How many candies are in 42 packs?

4. The distance between Houston and San Antonio is 200 miles. How many hours do I need to reach San Antonio from Houston if my average speed is 50 miles?

5. There are 369 pages in a book. How many hours will it take a boy to read it if he can read 41 pages in 1 hour?
6. The whole class went to see a play in a theater. How much did they pay for the tickets if one ticket costs $25 and there are 28 students in the class?

7. A group of high school students planted trees in a new park. They planted 168 trees altogether. How many students were there if each planted 3 trees?

8. After every trip a family makes a photo album with 48 pictures. How many trips have they made if they have 2688 pictures altogether?

9. A dentist pulls out 7 teeth every day. How many teeth does he pull out in one year, if he works 245 days a year?

10. During the average year I spend 3,285 hours sleeping. How many hours do I waste my time sleeping each day, if one year has 365 days?

11. An alien from Alpha Centauri has five heads. On each head, it has 26 ears. How many ears does it have altogether?

12. Jenny has 5 best friends. Each friend has 6 cats, and each cat has 7 kittens. How many cats and kittens are there altogether?

Chapter 2
Combined Operations and Mental Math

Lesson 1
Order of operations

You already know the four arithmetical operations: addition, subtraction, multiplication, and division. It is great, but not enough to solve problems involving more than one operation, such as 5 + 3 x 6. What should we do first: addition or multiplication? The answer depends on the order in which we do the operations. If we do the addition first, we get 48; if we do the multiplication first, we get 23. As you can see, there is quite a difference. This is why mathematicians have agreed on some rules about the order of operations. Without these rules, we could not be certain of getting the same answer for any problem. Here they are:

First, we should do multiplication and division, from left to right.
Next, we should do addition and subtraction, from left to right.

It means that in our example, we should first multiply 3 by 6 (we will get 18), and then add 5 and 18.
Here is how you write it: 5 + 3 x 6 = 5 + 18 = 23

Another example: 36 ÷ 4 - 2 x 4.
First do the division, then the multiplication, and finally the subtraction.

36 ÷ 4 - 2 x 4 = 9 – 2 x 4 = 9 – 8 = 1

Sometimes, however, we need to do the operations in a different order. Suppose we want to have a problem in which we do the addition before the multiplication; what should we do? In this case, we can use parentheses. All the operations inside the parentheses should be done first. But within the parentheses, we will follow the same order as before.

For example, let us do $(5+3) \times 6$.
First, add the numbers inside the parentheses, then multiply the result by 6.

$(5+3) \times 6 = 8 \times 6 = 48$ (compare this result with the result of the first example).

Now we can write the rules for the order of operations formally.

ORDER OF OPERATIONS

1. Do all the operations inside the parentheses.
2. Do all multiplication and division in order from left to right.
3. Do all addition and subtraction in order from left to right.

Here are some other examples.

$17 - 3 \times 5 = 17 - 15 = 2$ (multiply first, then subtract)

$7 \times (4+7) = 7 \times 11 = 77$ (compute what is inside the parentheses first, then multiply)

$14 \times 2 - 3 \times (6-4) = 14 \times 2 - 3 \times 2 = 28 - 6 = 22$ (first subtract the numbers inside the parentheses, then multiply 14 by 2, then 3 by 2, subtract last)

$36 \div (3+3) = 36 \div 6 = 6$ (add what is inside the parentheses first, then divide)

PRACTICE

1. $15 \div (3+2) =$
2. $15 \div 3 + 2 =$
3. $7 \times 3 + 6 \times 4 =$
4. $30 + 7 \times 8 =$
5. $7 \times (3+6) =$
6. $20 - 3 \times (5-2) =$
7. $18 \div 6 \times (4+2) =$
8. $11 + 2 \times (15 \div 3 - 2) =$
9. $6 \times (4 + 3 \times 2) \div (3+2) =$
10. $18 \div (2+4) \times 7 + 3 =$

Lesson 2
Rounding Whole Numbers

When we deal with large numbers, small differences become unimportant. Do you really care whether there are 996,978,197 grains of sand on a beach or 998,257,183? Or, for instance, do you really care, or even have any way of knowing, whether there are 11,045,897 or 11,045,923 people living in New York City? Since both of these numbers are unknown to us and likely to change over time, we prefer, instead, to **round** the number – to replace it by the closest number of billions or millions or hundreds, depending on what is important to us. So, we can say that the population of New York City is approximately 11 million, or 11,000,000; or that the number of grains of sand on a beach is approximately 1 billion, or 1,000,000,000. These numbers are much easier to deal with in calculations than the real numbers would be; and the difference is so small that it is unimportant.

Suppose we have 23,999,999 and we want to round it to the nearest million. What do you think, is this number closer to 24 million of to 23? If you said 24, you are right. Now, let us talk about how to round any number to any position. While the large numbers are more fun, we will start with smaller numbers, because they are easier.
Look at the number line.

0 1 2 3 4 5 6 7 8 9 10

5 is exactly in the middle of this interval. All the numbers that are bigger than 5 are closer to 10 than to 0, and all the numbers that are smaller than 5 are closer to 0 than to 10. You should always have this in mind when rounding numbers. Suppose you have to round 23 to the nearest ten. Definitely, 3 is closer to 0 than to 10; that is why we will round this number to 20 rather than to 30. But what about 28? You can see that this number is closer to 30 than to 20.

To see it better, let us draw another number line from 0 to 100.

0 10 20 30 40 50 60 70 80 90 100
5 15 25 35 45 55 65 75 85 95

You can see that every number with a number of units that is less than 5 is closer to the round number with the same amount of tens. At the same time, every number with a number of units that is greater that 5 is closer to the next amount of tens. It means that 42 is closer to 40, but 49 is closer to 50.
After seeing this number line, it is clear why we can use this rule for rounding two digit numbers:

if the number of units is less than 5, round down;
if the number of units is 5 or more, round up.

Let us round several two-digit numbers: $22 \approx 20$, $58 \approx 60$, $34 \approx 30$, $45 \approx 50$, $77 \approx 80$.
($\approx$ means "approximately equals").

Now I'll give you the rule of rounding much bigger numbers to any decimal place:

1. underline the digit in the place to which you want to round
2. look at the digit one place to its right:
if the digit is less than 5, round DOWN (the underlined digit will remain the same)
if the digit is 5 or more, round UP (the underlined digit will increase by one unit)
3. replace all the digits to the right of the underlined digit by zeros.

Let us look at an example.
Round 539 to the nearest hundred.
First step: underline the hundreds digit: $\underline{5}39$
Second step: look at the tens digit. It is 3, which is less than 5. Round down.
The result is 500. We will write it like this: $539 \approx 500$

Another example.
Round 45,722 to the nearest thousand.
First step: $4\underline{5},722$
Second step: $7 > 5$, round up.
Third step: $45{,}722 \approx 46{,}000$

If you know this rule very well and remember place values, it is not hard to round any number to any position. Let us round 324,561,777,932 to the nearest billion.
First step: $32\underline{4},561,777,932$
Second step: $5 = 5$, round up
Third step: $324{,}561{,}777{,}932 \approx 325{,}000{,}000{,}000$

Now try to do some problems yourself.

PRACTICE

Round to the nearest ten:

1. 58 ≈ 2. 33 ≈ 3. 81 ≈ 4. 99 ≈

Round to the nearest hundred:

5. 234 ≈ 6. 881 ≈ 7. 919 ≈ 8. 650 ≈

Round to the nearest thousand:

9. 5,623 ≈ 10. 36,199 ≈ 11. 9,911 ≈ 12. 287,455 ≈

Round to the nearest hundred thousand:

13. 222,895 ≈ 14. 7,573,583 ≈ 15. 92,152,743 ≈

Round to the nearest ten billion:

16. 25,749,832,902 ≈ 17. 454,912,012,350 ≈

Lesson 3

Mental Math: Estimation using rounding

As surprising as it may sound, the need to do calculations can sometimes arise outside of school, far away from pens and paper. We often have to add or subtract in a store, multiply or divide when we are in a car or do math somewhere else where we don't have anything but our brain. It is very important to be able to calculate fast. Sometimes you may only estimate the answer, but the result should be very close to the exact number.
There are a lot of ways to do it quickly and easily and to have a lot of fun along the way.

In the previous lesson, we talked about the cases when an approximate value of a number is more important that the exact value, and learned how to round numbers. It is very convenient to use rounded numbers when you need to estimate the sum or difference of some numbers quickly, especially when you have to work with large numbers.
Suppose 2 schools decided to merge into one. One school had 897 students, the second one had 1,231. Is the number of ones or even tens important? Of course not. The schools will round to the hundreds or maybe thousands before they will add these numbers and estimate the future enrollment.

897 ≈ 900

1,231 ≈ 1,200

900 + 1,200 = 2,100
Or they will round to the thousands:
897 ≈ 1,000
1,231 ≈ 1,000
1,000 + 1,000 = 2,000
In both cases the result shows that they should expect a little bit more than 2,000 students.
Compare this to the real answer.
897 + 1,231 = 2,128
As you see, the estimates are very close to the real answer, and take much less time.

When you go shopping, you might notice that the prices on the items often end in nines – $19, $39, or $499. This makes it difficult for you to add up the prices to figure out how much you should pay. However, if you estimate the total, it is much easier to figure out – it is easier to add 20+ 40 + 500 than it is to add the real numbers.
Rounding numbers can help you solve some hard problems in your head. Suppose you are in a department store and your mom has given you $50 to buy clothes. You choose a $19 pair of jeans, $18 sneakers, and a sweater that costs $25. Your mom would tell you at once that it is too much. How did she solve it so fast? She rounded the numbers. She added 20 + 20 + 30 and could see that it is more than 50. So, you have to find something cheaper.

Let us solve some problems using rounding.
Estimate the sum: 379 + 283.
Let us round both numbers to the nearest hundred:
379 ≈ 400,
283 ≈ 300.
Now it is very easy to estimate the sum: 400 + 300 = 700.

Estimate the difference: 32,888 - 25,199.
Let us round to the nearest thousand:
32,888 ≈ 33,000,
25,199 ≈ 25,000.
Now estimate the difference: 33,000 - 25,000 = 8,000.
You can see that after rounding numbers it is very easy to add or subtract the numbers mentally.

To use rounding for mental multiplication, we need to review multiplication by numbers with one or more zeros.
You remember that when we multiply a number by 10, we simply write 0 at the end of the number, like this: 35 x 10 = 350. We take 35 ten times, and we get 35 tens. You should also remember that if we multiply a number by 20 or 30, we should multiply the number by 2 or 3 and then write zero on the end of the product.
Actually, if we multiply numbers that end in zeros, we should begin by multiplying the nonzero portions of each number. Then we should write as many zeros as are at the end of both factors at the end of the product.
For example: 200 x 460. Multiply 2 x 46 = 92. Then write three zeros at the end.
Therefore, 200 x 460 = 92,000.
300 x 7000: 3 x 7 = 21 and write 5 zeros at the end: 300 x 7000 = 2,100,000
20 x 300 x 4,000: 2 x 3 x 4 = 24: 20 x 300 x 4,000 = 24,000,000

Now let us estimate a product using rounding.

39 x 42 ≈ 40 x 40 = 1600
217 x 589 ≈ 200 x 600 = 120,000

We can use rounding for estimating a quotient, too.
At first, let us talk about dividing numbers with zeros at the end. For example:
1600 ÷ 400. You are dividing 16 hundred by 4 hundred, which is the same thing as dividing 16 by 4, and the answer is 4.
350 ÷ 70 is the same as 35 tens divided by 7 tens, and the answer is 7.
Therefore, we can cross out the same number of zeros at the end of the dividend and the divisor, which will make calculations much easier.
399 ÷ 42 ≈ 400 ÷ 40 = 40 ÷ 4 =10
5, 127 ÷ 48 ≈ 5,000 ÷ 50 = 500 ÷ 5 = 100

PRACTICE.

Estimate by rounding. Round so you can use mental math.

98 + 33 ≈ 127 + 61 ≈ 77 - 28 ≈ 1,293 - 689 ≈ 96 + 97≈

12,677 + 5,286 ≈ 455 - 279 ≈ 519 + 324 ≈ 1,766 – 812 ≈

309 + 592 + 273 ≈ 893 - 283 - 311 ≈ 1,685 + 309 - 1,088 ≈

Use mental math to find the products.

200 x 70 = 30 x 40 = 900 x 500 = 40 x 50 x 60 x10 =

20 x 30 x 50 = 6,000 x 7,000 = 300 x 400 x 500 x 100 =

Estimate by rounding. Round so you can use mental math.

49 x 9 ≈ 716 x 28 ≈ 514 ÷ 25 ≈ 99 x 199 ≈ 1,987 ÷ 2,006 ≈

307 x 287 ≈ 52 x 21 x 39 ≈ 19 x 28 x 97 ≈ 189 ÷ 93 ≈

Lesson 4
Mental Math: Front – End Estimation

Front-end estimation helps us approximate sums or differences of numbers quickly.
Suppose we want to add 384 and 235. At first we will add the digits with the greatest place value: 3 + 2 = 5. Therefore, we already have 500. This is a rough estimate – we know that the answer is around 500. Then, we will use the rest of the digits to adjust the estimate: 84 + 35 will give us a little bit more than 100.
As a result, we can say that 384 + 235 ≈ 600.

As you see, front-end estimation is a method that requires 2 steps:

1) to get a rough estimate we add the digits with the greatest place value;
2) then we use the other digits to adjust the estimate.

Front-end estimation will give you an approximate answer. If you are in a hurry, and don't care about accuracy, this will be good enough; but it won't be the right answer.
For better understanding, try to solve several problems using this table:

TARGET 50		
28	35	43
25	6	14

Your job is to find pairs of numbers from this table whose sum is close to 50. What does "close" mean? Numbers between 45 and 55 are close enough. Let us start with 28. Look at its neighbor 35. If we add 2 tens and 3 tens, we get 5 tens, but after we adjust, we will get a number greater than 10. That means that our sum is greater than 60; this pair of numbers does not work. You can see at once that 28 + 43 gives us a sum greater than 50 (if we add 2 tens and 4 tens, we get 6 tens); so, let us look at the next number, which is 25. 2 + 2 = 4, giving us 4 tens, and 8 + 5 is close to 10. This pair works! Try to find two other pairs yourself.

Now we will solve several problems using front-end estimation and decide whether the exact sum or difference is more or less than our estimate.

321 + 573: 3 + 5 = 8 (we have 800), 21 + 73 ≈ 100, therefore 321 + 573 ≈ 800+100=900

$$\begin{array}{r} 321 \\ +\underline{573} \\ 894 \end{array}$$

The exact value is a little bit less than the estimate, but very close.

742 - 238: 7 - 2 = 5 (we have 500), 42 - 38 ≈ 0, therefore 742 - 238 ≈ 500

$$\begin{array}{r} 742 \\ -\underline{238} \\ 504 \end{array}$$

The exact value is a little bit more than the estimate, but very close.

PRACTICE

Use front-end digits to estimate:

657 + 261 ≈ 837 - 529 ≈ 495 + 596 ≈ 693 – 105 ≈

5,678 - 1,596 ≈ 987 + 210 + 499 ≈ 1,203 + 2,611 + 3,222 ≈

Estimate to find the best missing number for each ?

Missing numbers: 13, 17, 42
1) 39 + ? is close to 80
2) 52 - ? is close to 40
3) 84 + ? is close to 100

Lesson 5
Mental Math: Estimation Using Compensation

Before we begin estimation using compensation, let us talk about compatible numbers. **Compatible** numbers are numbers that are easy to compute. For example:
75 + 25 = 100
25 x 4 = 100
68 - 18 = 50
8,000 / 8 = 1,000
You can find the answers in your head.

There are several techniques that can make mental calculations easier. Let us start with **compensation.** This is a very good way to calculate. For example, suppose you need to find the sum of 99 and 137. Add 1 to 99 and you will have 100. It is extremely easy to add 100 to any number, 137 + 100 = 237. Now you can subtract 1 from this sum, which is not hard either: 237 - 1 = 236. Congratulations! You just did a complicated calculation in your head.

I think that you understand that compensation is a method in which we change a number to make it easier to work with, and adjust the answer afterwards. This method will give you the exact answer.

Let us do some more calculations using compensation.

36 + 8: 1) 36 + 10 = 46 2) 46 - 2 = 44

374 + 297: 1) 374 + 300 = 674 2) 674 - 3 = 671

67 - 29: 1) 67 - 30 = 37 2) 37 + 1 = 38

25 x 99: 1) 25 x 100 = 2500 2) 2500 - 25 = 2475

PRACTICE

49 + 37 = 456 - 9 = 37 x 9 = 5,842 + 1,998 = 671 – 199=

56 x 999 = 497 + 358 = 3,762 - 2,998 = 675 – 99 =

Lesson 6

Even More Mental Math (substituting compatible numbers and breaking apart numbers)

If you need to estimate a sum and want to do it mentally, it is very convenient to use a technique called substituting compatible numbers. What does it mean and how do we work with it?

Let us begin by looking at an example.
Suppose you want to add 287 and 319. Replace these numbers with numbers that are easy to compute mentally; then you can add them easily.

Replace 287 + 319 with 275 + 325. Now add: 275 + 325 = 600
If you like, you can add 287 and 319 and you will see that your estimate is very close to the exact answer.

Another example: 458 - 229. Replace it with 450 - 225. Now subtract: 450 - 225 = 225

You see that this method requires replacing some or all of the numbers with numbers that are easy to compute (we call them compatible numbers) and then you can make your calculations mentally. With this method, you will not get the exact answer, but rather an estimate.

Try to solve these problems yourself and compare the answers.

$179 + 118 \approx 175 + 125 = 300$ $83 - 27 \approx 85 - 25 = 60$

$686 - 293 \approx 700 - 300 = 400$ $73 + 56 + 28 + 141 \approx 75 + 50 + 25 + 150 = 300$

We can use this method for multiplication, too.

For example: $26 \times 4 \approx 25 \times 4 = 100$
$69 \times 11 \approx 70 \times 10 = 700$

Now let us talk about breaking numbers apart.
Any number with more than 1 digit can be broken apart. For example: 28 = 20 + 8, 349 = 300 + 40 + 9. You should remember place values, and then you can do it easily. Sometimes, we need this method to solve math problems mentally. This will give us an exact answer, not an estimate.

Suppose we want to multiply 47 by 6. Actually, this is the same as multiplying 40 by 6, 7 by 6 and then adding the products. We can write it like this: (40 x 6) + (7 x 6) = 240 + 42 = 282.
It is called the **distributive property** of multiplication over addition.
Let us solve several problems together.

2 x 64 = 2 x 60 + 2 x 4 = 120 + 8 = 128

3 x 47 = 3 x 40 + 3 x 7 = 120 + 21 = 141

108 x 6 = 100 x 6 + 8 x 6 = 600 + 48 = 648

It is not hard, because you already know how to multiply numbers with zeros at the end.

PRACTICE

Substitute compatible numbers to estimate an answer.

$247 - 58 \approx$	$178 + 126 \approx$	$888 - 493 \approx$	$1{,}987 + 2{,}072 \approx$
$27 \times 9 \approx$	$629 \div 69 \approx$	$26 \times 98 \approx$	$428 + 377 \approx$

Use the distributive property and mental math to find the products.

3 x 28 =	4 x 17 =	5 x 43 =	58 x 2 =	102 x 3 =
204 x 4 =	8 x 15 =	6 x 43 =	77 x 3 =	504 x 2 =

Lesson 7
Word Problems

Estimate; try to solve the problem in your head.

1. Mr. Smith bought an old car for \$6,200. The first year he paid \$840 for repairs, the second year he spent \$1,920 for repairs, and the third year he paid \$912. How much did the car really cost him?

2. The 5th grade class is going to the theater. How much will they pay for tickets if each ticket costs \$29 and there are 32 students in the class?

3. The distance between San Antonio and Houston is 197 miles. How many hours will it take you to reach Houston from San Antonio if your average speed is 48 miles per hour?

4. A typist can type 93 words per minute. How many words will the typist type in an hour?

5. A woman had \$5,673 in her bank account. How much will she have after she spends \$1,999?

6. A school has 192 students. How many tables should they have in the lunch room if only 9 people can sit at each table?

7. A family sold their old house for $82,500 and wants to buy a new house for $133,000. How much money should they add to buy the new house?

8. A young couple sent out 92 invitations to their wedding. Each invitation costs $8. How much money have they spent on invitations?

9. A school bought 42 math textbooks and paid $1638 for them. How much does each textbook cost?

10. A club is going to buy 24 jump ropes. Each rope costs $7. How much money will the club spend?

Chapter 3
Number Theory

Lesson 1
Factors and Multiples

You already know that when we multiply two numbers, the numbers that are multiplied are called **factors**, and the result of the multiplication is called the **product**.

$3 \times 8 = 24$

In this example, 3 and 8 are the factors of 24, and 24 is the product.

If you think about it, you will be able to find other numbers that, when multiplied, form the same product. For example, $4 \times 6 = 24$. In many situations, it is important to know all the factors of a number. How can we find them?

To do this, we will use division. If you want to know whether or not one number is a factor of another one, divide the second number by the first one. If you get a whole quotient without a remainder, then the divisor is a factor of the dividend.

Suppose you want to find out whether or not 7 is a factor of 126. $\frac{126}{7} = 18$, therefore, $18 \times 7 = 126$ and both 18 and 7 are factors of 126.

To find all the factors of a whole number, divide the number by each natural number. When the factors start repeating, you can stop; you have found all of them.

Let us find all the factors of 18.

$18 \div 1 = 18$; 1 and 18 are factors of 18

$18 \div 2 = 9$; 2 and 9 are factors of 18

18 ÷ 3 = 6; 3 and 6 are factors of 18
18 ÷ 4 = 4 R2; 4 is not a factor of 18
18 ÷ 5 = 3 R3; 5 is not a factor of 18
18 ÷ 6 = 3; no new factors appear, therefore, we can stop.
So, the factors of 18 are 1, 2, 3, 6, 9, and 12.

To make this process easier, we can draw something like a rainbow. Each arc connects two factors that go together.

The factors of 18: 1, 2, 3, 6, 9, 18

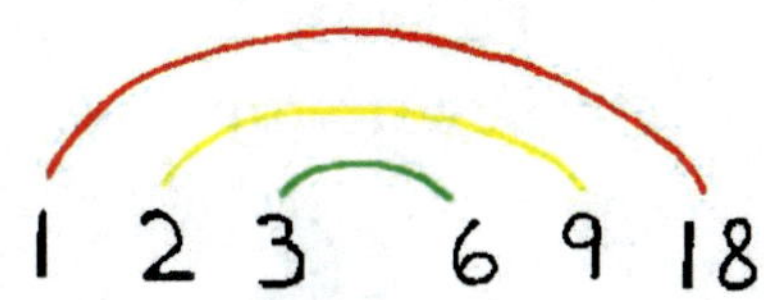

The factors of 24: 1, 2, 3, 4, 6, 8, 12, 24

The product of two whole numbers is a **multiple** of each of these two numbers.
15 = 3 x 5, so 15 is a multiple of 3, as well as a multiple of 5. It is very easy to find multiples of a number; just multiply the number by 0, 1, 2, 3, and so on.

5 x 0 = 0   5 x 1 = 5   5 x 2 = 10   5 x 3 = 15   5 x 4 = 20   5 x 5 = 25…

And so 0, 5, 10, 15, 20, 25,… are all multiples of 5. There are an infinite number of multiples, because there are an infinite number of whole numbers.

Now, we can come up with a new definition of even and odd numbers. All multiples of 2 are even numbers, and all the other whole numbers are odd numbers.

Let us find the first seven nonzero multiples of 8. To do that, simply multiply 8 by 1, 2, 3, 4, 5, 6, and 7. You will get 8, 16, 24, 32, 40, 48, 56.
See? It's not difficult at all.

Now, practice.

PRACTICE

Is the first number a factor of the second? Answer yes or no.

1) 5, 35   2) 7, 46   3) 8, 60   4) 2, 124   5) 9, 81   6) 12, 144

Find all the factors of the given number.

1) 32 2) 50 3) 64 4) 100 5) 53 6) 78 7) 67

State the first five nonzero multiples of each number.

1) 3 2) 7 3) 11 4) 15 5) 23 6) 33 7) 10

Lesson 2
Divisibility

To find the factors of a number, you have to divide the number by a lot of numbers, and find the factor only when you get a quotient without a remainder. This takes a long time; and so, it is very convenient to be able to see at once whether or not you have a remainder. In this lesson, I will show you how to do it.

A whole number is **divisible** by another whole number if the quotient of these two numbers is a whole number and there is no remainder (or we can say that the remainder is 0). 28 is divisible by 7 because $28 \div 7 = 4$ with no remainder. A whole number, obviously, is divisible by each of its factors.

For some factors, there is an easy way to determine divisibility without actually dividing. This can be very helpful if the dividend is large. Here are some tricks to help you.

I think that you can see that **every number is divisible by 1**, because every number has 1 as a factor.

Every even number is divisible by 2.

$12 \div 2 = 6$ $26 \div 2 = 13$ $38 \div 2 = 19$ $60 \div 2 = 30$

A number is divisible by 3 if the sum of its digits is divisible by 3. (I won't show you why it works, as the proof is still a bit hard for you; you just have to trust me for now).

42 is divisible by 3, because $4 + 2 = 6$ and 6 is divisible by 3. $42 \div 3 = 14$

711 is divisible by 3, because $7 + 1 + 1 = 9$ and 9 is divisible by 3. $711 \div 3 = 237$.

A number is divisible by 4 if its last two digits are divisible by 4. This works because 100 is divisible by 4 (100 ÷ 4 = 25); so any whole number of hundreds is going to be divisible by 4. Any remainder would have to come from the tens and ones, or from the last two digits.
732 is divisible by 4, because 32 is divisible by 4. 732 ÷ 4 = 183
24,643 is not divisible by 4, because 43 is not divisible by 4.

A number is divisible by 5 if its ones digit is 0 or 5. This is easy to see; because 10 is divisible by 5, any whole number of tens is also going to be divisible by 5. The only number of units that is divisible by 5 is 5. So, we can only get a ones digit of 0 or 5 if the number is going to be divisible by 5.
825 is divisible by 5, because its ones digit is 5. 825 ÷ 5 = 165.
670 is divisible by 5, because its ones digit is 0. 670 ÷ 5 = 134.
99 is not divisible by 5, because its ones digit is 9.

A number is divisible by 6 if it is divisible by 2 and by 3. This works because 6 = 2 x 3.
732 is divisible by 6, because it is an even number (and thus divisible by 2) and the sum of its digits is 12 which is divisible by 3. 732 ÷ 6 = 122
711 is not divisible by 6 because it is an odd number.

A number is divisible by 9 if the sum of its digits is divisible by 9. The way this works is still a bit hard for you, so I will explain it later.
342 is divisible by 9, because 3 + 4 + 2 = 9. 342 ÷ 9 = 38.
1,234 is not divisible by 9, because 1 + 2 + 3 + 4 = 10, but 10 is not divisible by 9.

A number is divisible by 10 if its ones digit is 0. This should be obvious.
940 is divisible by 10, because its ones digit is 0.
682 is not divisible by 10, because its ones digit is 2, not 0.

There aren't any good tricks for other numbers, so you'll have to divide using mental math or use long or short division to see whether or not a number is divisible by another number.

PRACTICE

State whether or not the number is divisible by 2.
1) 46 2) 39 3) 102 4) 18 5) 111 6) 1,234

State whether or not the number is divisible by 3.
1) 87 2) 111 3) 122 4) 2,341 5) 45,561 6) 5,666

State whether or not the number is divisible by 4.
1) 164 2) 672 3) 3,411 4) 67,232 5) 123,456 6) 1,000

State whether or not the number is divisible by 5.
1) 85 2) 102 3) 1,292 4) 10, 000 5) 55, 553 6) 100, 000

State whether or not the number is divisible by 6.
1) 663 2) 102 3) 2, 194 4) 99 5) 624, 000 6) 326

State whether or not the number is divisible by 9.
1) 1,260 2) 668 3) 298,011 4) 5,769 5) 786 6) 111,111

State whether or not the number is divisible by 10.
1) 1,260 2) 23,705 3) 600 4) 89,999 5) 19,800 6) 40,000,001

Lesson 3
Prime and Composite Numbers

Perhaps you noticed that any number can be divided evenly by some other number.
Some numbers can only be divided by 1 or by the number itself; for example, 3 or 5. Some numbers, however, can be divided by other numbers as well; for example, 6 can be divided by 2, and 15 can be divided by 3. The first kind of numbers are called prime numbers; the second kind are called composite numbers. In summary, a number which has exactly two factors – 1 and the number itself – is called a **prime number.** A number that has more than two factors is called a **composite number**. Since 1 has exactly 1 factor, it is neither prime nor composite.

It is easy to figure out whether 3 or 7 are prime or composite, but how do we tell if a really large number is prime or composite? That question has occupied mathematicians for many years; and as early as about 230 BC, Eratosthenes, a great Greek mathematician, found a very clever way to separate prime and composite numbers. Here is how he did it.
Write all natural numbers from 2 to a certain maximum number, then cross out all the multiples of 2 except 2 itself. Next, cross out all the multiples of 3, except 3 itself. Continue crossing out multiples of each remaining number except for the number itself. The numbers that remain at the end of this process are the prime numbers.

This method is called the **sieve of Eratosthenes.**

~~1~~ **2** **3** ~~4~~ **5** ~~6~~ **7** ~~8~~ ~~9~~ ~~10~~ **11** ~~12~~ **13** ~~14~~ ~~15~~ ~~16~~ **17** ~~18~~ **19** ~~20~~

~~21~~ ~~22~~ **23** ~~24~~ ~~25~~ ~~26~~ ~~27~~ ~~28~~ **29** ~~30~~ **31** ~~32~~ ~~33~~ ~~34~~ ~~35~~ ~~36~~ **37** ~~38~~

~~39~~ ~~40~~ **41** ~~42~~ **43** ~~44~~ ~~45~~ ~~46~~ **47** ~~48~~ ~~49~~ ~~50~~

You can see that the numbers 2, 3, 5, 7, 11, 13, 17, 19, 23, 29, 31,37, 41, 43, and 47 are prime. It is very useful to remember at least this many prime numbers; if you want to know more primes, you can start out with a longer list of natural numbers and apply the same method.

You can see that the only even prime number is 2; all the other even numbers can be written as a product of 2 and another number. For example, 8 = 2 x 4, 18 = 2x9, 26 = 2x13, etc.

PRACTICE

State whether each number is prime or composite.

1) 4 2) 11 3) 15 4) 23 5) 27 6) 45 7) 53

8) 57 9) 69 10) 71 11) 81 12) 95 13) 99 14) 1

Lesson 4
Prime Factorization

Each number can be written as a product of 2 or more numbers, and sometimes can be written in several different ways. For example, 24 = 3x8 = 2x12 = 4x6. If we write a number as a product of prime numbers, it is called the **prime factorization** of that number. This is important because there is only one way to write a number as a product of prime numbers; the prime factorization defines the number.

The easiest way to figure out the prime factorization is by making a factor tree. Here is how you do it. At first, write the number as a product of 2 factors- it doesn't matter which factors you choose. Then, look at the factors that are composite and write them as products of 2 factors; continue this until there are no composite factors left.

Here is how you write it:

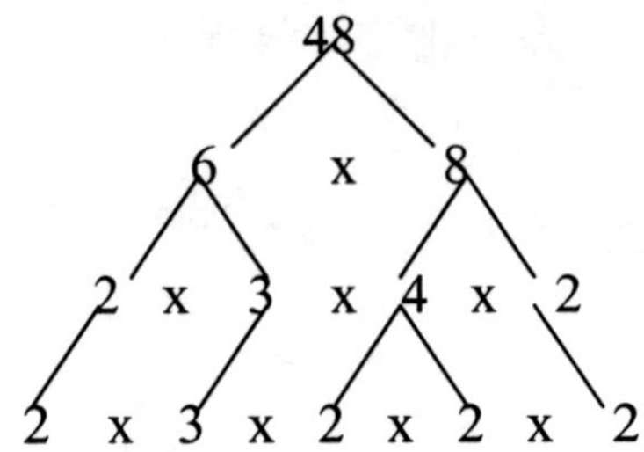

So, the prime factorization of 48 is 2x2x2x2x3.
Note that if you started with a different pair of factors, for instance 16x3, you would still get the same prime factorization. That's why we can say that **the prime factorization is unique**. Here is another example.

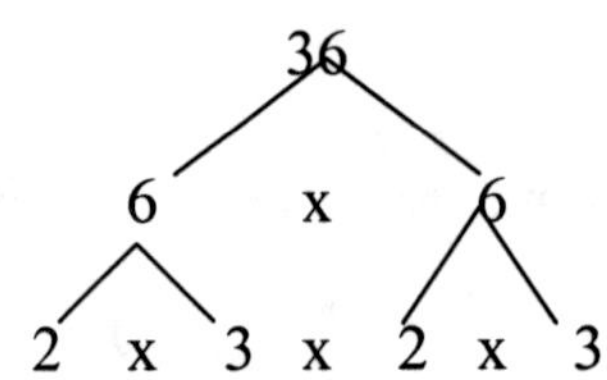

So, 36 = 2x2x3x3. Because all the factors here are prime, we can say that this is the prime factorization.

There is another way to find the prime factorization of a number; it is called the "ladder method". For this method, we will use repeated division. Beginning with 2, check whether each prime is a factor of the number. If it is, divide the number by the prime. Continue until you get a quotient of 1. The product of the divisors is the prime factorization of the number.

```
2 |_50
   5 |_25
      5 |_5
         |_1
```

50 = 2 x 5 x 5

Look at these examples.

2 | 84
2 | 42
3 | 21
7 | 7
| 1

$84 = 2 \times 2 \times 3 \times 7$

2 | 100
2 | 50
5 | 25
5 | 5
| 1

$100 = 2 \times 2 \times 5 \times 5$

2 | 162
3 | 81
3 | 27
3 | 9
3 | 3
| 1

$162 = 2 \times 3 \times 3 \times 3 \times 3$

Now practice.

PRACTICE

Make a factor tree to find the prime factorization of each number.

1) 48 2) 28 3) 30 4) 100 5) 120

Use repeated division to find the prime factorization of each number.

6) 108 7) 96 8) 225 9) 144 10) 297

Lesson 5
The Greatest Common Factor

The **Greatest Common Factor** of two whole numbers is the greatest whole number that is a factor of both numbers. It is so important and used so often that mathematicians use a special abbreviation for greatest common factor: **GCF**.
Let us find the GCF for two numbers: 24 and 40.
At first we will make a list of factors for both numbers.
24 : 1, 2, 3, 4, 6, 8, 12, 24 40 : 1, 2, 4, 5, 8, 10, 20, 40
Then we will find all common factors for both numbers.
1, 2, 4, 8
Now you can see that the Greatest Common Factor for 24 and 40 is 8.

To find the GCF for two numbers it is not necessary to make a list of factors for both numbers. You can make a list of factors only for the smallest number and then check each factor to see whether or not it is also a factor of the second number. Then, you can choose the greatest one of them.

For example, let us find the GCF for 32 and 80.
Make a list of factors for 32: 1, 2, 4, 8, 16, 32

Now check these factors to see whether or not they are factors of 80. Because we are looking for the greatest common factor, it is reasonable to start from the greatest factor and go down.
32 is not a factor of 80; 16 is a factor of 80. Therefore, 16 is the GCF for 32 and 80.
It is not necessary to check other factors, because even if they are common factors for 80, they are smaller than 16.
Another example. Find the GCF for 50 and 60.
50 : 1, 2, 5, 10, 25, 50
Neither 50 nor 25 is a factor of 60; 10 is a factor of 60.
So, 10 is the GCF of 50 and 60.

Sometimes the smaller number is a factor of the larger one. In this case this number is the GCF of both numbers. For example, look at 12 and 36. 12 is a factor of 36, therefore 12 is the GCF of 12 and 36. This is why it is a good idea to check whether the smaller number is a factor of the larger one or not.

There is another way to find the GCF of two numbers. In this way we will get the prime factorizations of both numbers and then find the product of all the common factors. This product will be a common factor because it is made up of all the prime factors that both numbers have.
Look at the example. Suppose we are looking for the GCF of 36 and 48.
We will give the prime factorizations for 36 and for 48, then underline all the common factors for both numbers, and after that we will multiply them. It is like building our common factor out of building blocks.

$36 = \underline{2} \times \underline{2} \times \underline{3} \times 3$ $\qquad$ $48 = \underline{2} \times \underline{2} \times 2 \times 2 \times \underline{3}$

$\text{GCF} = 2 \times 2 \times 3 = 12$

Let us find the GCF for 30 and 45.

$30 = 2 \times \underline{3} \times \underline{5}$ $\qquad$ $45 = \underline{3} \times 3 \times \underline{5}$

$\text{GCF} = 3 \times 5 = 15$

Sometimes, two numbers don't have any common factors but 1. They are called **relatively prime**. For example: 4 and 5; 8 and 9; 25 and 32 and so on.

We can find the GCF not only for two, but for three and even more numbers. The methods are the same. For example, find the GCF for 30, 36, 42
Let us use the method with prime factorization.
$30 = \underline{2} \times \underline{3} \times 5$ $\qquad$ $36 = \underline{2} \times 2 \times \underline{3} \times 3$ $\qquad$ $42 = \underline{2} \times \underline{3} \times 7$

$\text{GCF} = 2 \times 3 = 6$

PRACTICE

List the factors of both numbers to find the GCF of each pair.
1) 6, 8 2) 24, 28 3) 18, 30

List the factors of the smaller number to find the GCF of each pair.

4) 18, 27 5) 12, 40 6) 56, 80

Use prime factorization to find the GCF of each pair.

7) 45, 60 8) 36, 54 9) 144, 180

Use any method to find the GCF of each set of numbers.

10) 32, 48, 80 11) 40, 50, 60

Lesson 6
The Least Common Multiple

The **Least Common Multiple** of two whole numbers is the smallest nonzero whole number that is a multiple of both of the whole numbers.
Again we have an abbreviation for this very important number: **LCM**.
You already know how to find multiples for any number, so it will not be hard to find multiples for two numbers. Then, if we need the LCM for these two numbers, we will find the smallest common multiple of the numbers.
For example, let us find the LCM of 6 and 9.
Multiples of 6: 0, 6, 12, 18, 24, 30, 36, . . .
Multiples of 9: 0, 9, 18, 27, 36, 45, . . .
Common Multiples of 6 and 9: 0, 18, 36, . . .
Least Common Multiple of 6 and 9: 18

To make finding the LCM easier, we can list only the multiples of the larger number, then choose the smallest one that is also a multiple of the smaller number.
For example, let us find the LCM of 18 and 30.
At first we will list multiples of 30:
0, 30, 60, 90, . . .
We can see that 90 is a multiple of 18, and because neither 30 nor 60 is a multiple of 18, 90 is the LCM of 18 and 30.

We can use prime factorization to find the LCM of two numbers. At first we should write the prime factorizations of both numbers and underline all the common factors for both numbers. Then write the prime factorization of one of the numbers and include factors from the second that have not been included in the factors you wrote out. It means that you will include factors from the second number which are not underlined.
For example: find the LCM of 12 and 20.

$12 = \underline{2} \times \underline{2} \times 3$ $\quad 20 = \underline{2} \times \underline{2} \times 5$

LCM (12, 20) = 2 x 2 x 3 x 5 = **60**

Another example: find the LCM of 28 and 40.

28 = 2 x 2 x 7 40 = 2 x 2 x 2 x 5

LCM (28, 40) = 2 x 2 x 7 x 2 x 5 = **280**

You can see that each number is a factor of the LCM: 280 = 28 x 10
280 = 40 x 7

Sometimes the larger number is a multiple of the smaller one; in this case we can tell at once that the larger number is the LCM of the numbers.
For example: find the LCM of 7 and 21. 21 is a multiple of 7 (7 x 3 = 21), therefore 21 is the LCM of 7 and 21.

If two numbers are relatively prime, it is very easy to find their LCM. Multiply them and the product is the LCM of the numbers.
For example, find the LCM of 8 and 9.
They are relatively prime, so their LCM is 8 x 9 = 72

We can find LCM not only for two numbers, but for three and more numbers. The algorithm will be the same.
Let us find the LCM of three numbers: 9, 12, 15
At first we will write the prime factorization of all three numbers:
9 = 3 x 3 12 = 2 x 2 x 3 15 = 3 x 5
LCM (9, 12, 15) = 3 x 3 x 2 x 2 x 5 = 180
To find the LCM of 9, 12, 15 we wrote the prime factorization of 9: 3 x 3, then included 2 x 2 from the prime factorization of 12 (3 was there as a factor of 9), then included 5 from the prime factorization of 15 (3 was there as a factor of 9).

PRACTICE

List the multiples of the numbers to find the LCM of each pair.
1) 5, 6 2) 4, 9 3) 12, 15

List the multiples of the larger number to find the LCM of each pair.
4) 8, 12 5) 5, 18 6) 3, 17

Use prime factorization to find the LCM of each pair.
7) 15, 20 8) 40, 50 9) 27, 36

Find the LCM of the numbers given using any method.
10) 9, 45 11) 14, 15 12) 30, 36, 42

Lesson 7
Word Problems

1. A basketball coach has to schedule training sessions for 18 boys and 24 girls with the same number of kids in each session. What is the largest training session size possible?

2. A family wants to tile a living room with square pieces of tile that are all the same size. What is the largest square piece can be used if the room has a rectangular shape with dimensions 36 feet by 66 feet?

3. Mary and Jake need pieces of ribbon 12 or 16 cm long for a school project. Because they are still not sure about the length, they decided to order such a length of ribbon as to be able to cut it either into a whole number of 12-cm pieces or into a whole number of 16-cm pieces. What is the shortest length of ribbon that they can order?

4. To build a house a builder used two kinds of bricks, with heights 15 and 18 cm respectively. What is the least number of rows of each kind that will have equal heights?

5. Find two whole numbers whose product is 96 and whose LCM is 24.

6. Find two whole numbers whose product is 54 and whose GCF is 3.

7. There are 12 fifth graders and 16 sixth graders in a science class. A teacher wants them to work in teams with the same number of students in each team. What is the largest number of students in each team possible?

8. One alarm clock was set to go off every 6 hours, and another one was set to go off every 8 hours. How many times per day will they go off at the same time?

9. Two families travel together in two cars. If one car gets 20 miles per gallon and another gets 24 , what is the smallest whole number of gallons of gasoline each car need to travel exactly the same distance as the other car?

10. Tom wants to cover a kitchen floor with tiles. His kitchen is 120 inches long and 90 inches wide. In the store he can buy tiles with size 12 x 12 , 8 x 8, or 6 x 6. If he does not want to cut the tiles, which size should he choose?

Chapter 4
Fractions and Mixed Numbers

Lesson 1
Part of a Whole

In our everyday life, we often have to divide an object into several equal parts. For example, you can't eat pizza before slicing it into 4 or even 8 parts.

If you have only one apple and want to share it with your best friend, you will divide it into 2 equal parts. When your mom buys a chocolate bar for you, she might ask you not to eat it at once, but to cut it into 3 pieces and eat only one piece today.

In the language of mathematics, these pieces are represented by fractions.

Look at this rectangle:

Now let us divide it into 2, 3, and 4 equal parts.

2 equal parts 3 equal parts 4 equal parts

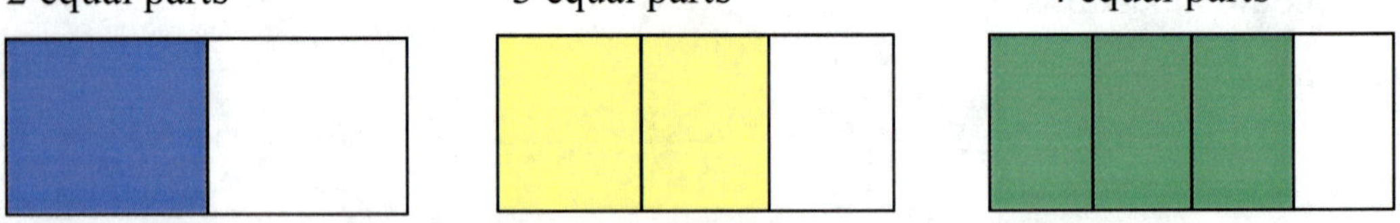

2 halves 3 thirds 4 fourths

We write a fraction as number with 2 levels: the number on top is called the **numerator**, and the number on the bottom is called the **denominator**. The **denominator** shows us into how many equal parts a whole number was divided, while the **numerator** shows us how many of these parts we have.

Now let us write fractions to name the shaded part of the rectangles above.

1) $\frac{1}{2}$ 2) $\frac{2}{3}$ 3) $\frac{3}{4}$

To read a fraction, say the top number as a cardinal number, and then say the bottom number as an ordinal number. If you don't know the difference between cardinal and ordinal numbers, just remember that an ordinal number shows the order of a number in a sequence, like first, second, third. A cardinal number is a regular counting number, like one, two, three.

Let us practice: $\frac{1}{3}$ is one third, $\frac{3}{4}$ is three fourths, $\frac{1}{5}$ is one fifth, $\frac{2}{7}$ is two sevenths.

Now, let us look at the pictures and write a fraction to name the shaded part. We will write the fractions as a number and with words.

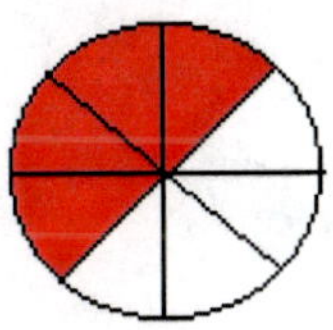

$\frac{4}{8}$
four eighths

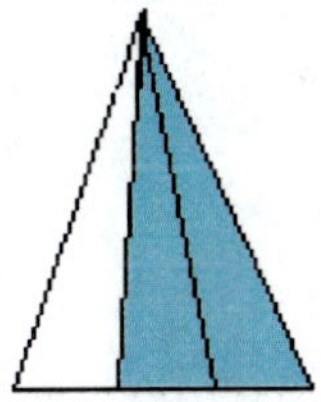

$\frac{2}{3}$
two thirds

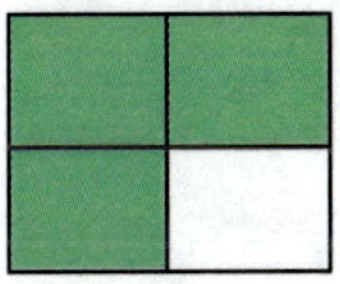

$\frac{3}{4}$
three fourths

PRACTICE

Give the fraction of the region that is colored. Write the fraction as a number and with words.

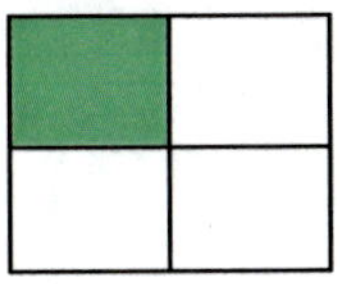

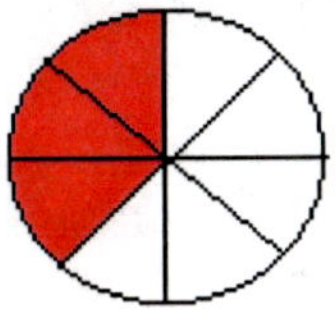

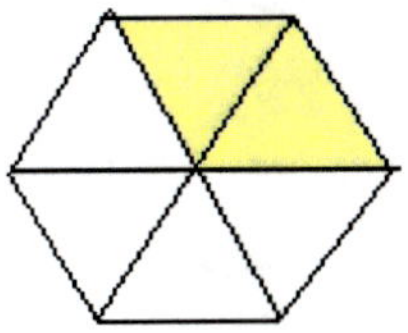

Write the fractions.

one third two eighths four sevenths one ninth

Color in $\frac{7}{8}$ of the figure:

Color in $\frac{4}{5}$ of the figure:

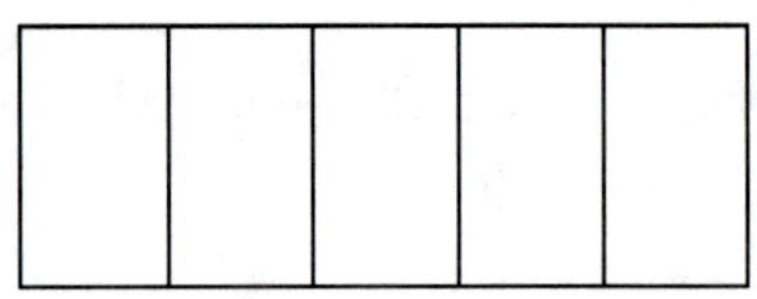

Lesson 2
Part of a Set

In the last lesson, we produced fractions by dividing one object - one pizza or one apple - into several pieces and taking some of these pieces. However, the object we start with does not have to be just one apple or one pizza – it could be a set of several objects. For example, we could divide a set of 5 crayons into five equal parts – each part is one crayon. Then, we could say that $\frac{3}{5}$ of that set is 3 crayons.

Suppose a class of 25 students had a test, and these are the results: 7 students got A's, 12 got B's, and 6 got C's. We can write that $\frac{7}{25}$ of the class got A's, $\frac{12}{25}$ of the class got B's, and $\frac{6}{25}$ of the class got C's.

For your birthday party your mom bought 12 balloons: 3 green ones, 4 blue ones, and 5 red ones. You took all the green balloons to decorate a porch. What fraction will show what part of the balloons you used?

The set of balloons was already divided into 12 equal parts, and so we will have 12 in the denominator. You took only 3 balloons, so we have 3 in the numerator.

$$\frac{\text{number of green balloons}}{\text{number of balloons in all}} \qquad \frac{3}{12}$$

I think that you understand that the fraction $\frac{4}{12}$ shows what part of the balloons are blue, and the fraction $\frac{5}{12}$ shows what part of the balloons are red.

What fraction would you use for one balloon? $\left(\frac{1}{12}\right)$

What fraction would you use for all the balloons? $\left(\frac{12}{12}\right)$

Now, if you understand what fractions are, it is not so difficult to find a fractional part of a number. I will repeat that the denominator shows by how many parts we divide the number, and the numerator shows how many of these parts we have.

Suppose we want to find $\frac{1}{3}$ of 24. If we divide 24 by 3, we will find exactly one third of 24; it is 8. To find $\frac{2}{3}$ of 24, we should find one third of 24 and then multiply it by 2, because we need to find two thirds of 24. This is the way to find it: $\left(\frac{24}{3}\right)\times 2 = 16$.

Let us solve several problems together:

$\frac{1}{5}$ of 25 $= \frac{25}{5} = 5$

$\frac{2}{7}$ of $49 = \left(\frac{49}{7}\right) \times 2 = 14$

$\frac{3}{4}$ of $40 = \left(\frac{40}{4}\right) \times 3 = 30$

$\frac{1}{8}$ of $64 = \frac{64}{8} = 8$

Now solve some problems yourself.

PRACTICE

What fraction represents the red part of each group?

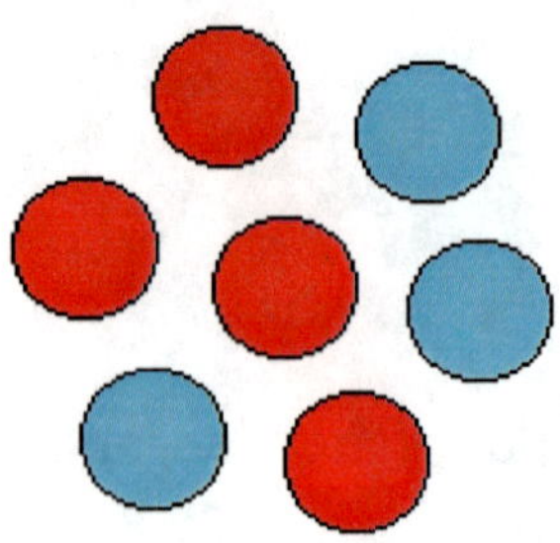

__________ __________

Give the fraction of each set that is colored.

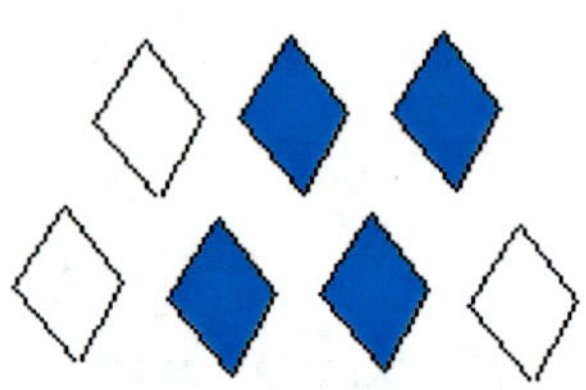

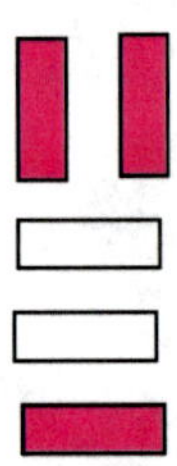

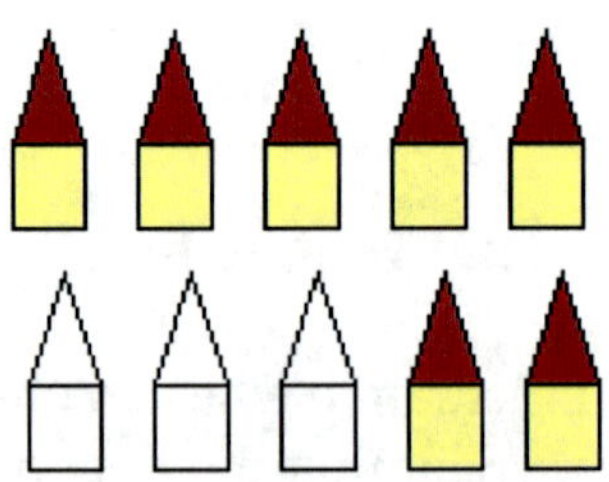

________ ________ ________

Answer the questions:
There are 13 boys in a class. 4 of them play soccer, 3 boys study karate, and 5 like to swim.

a) What fraction of the boys play soccer?
b) What fraction of the group are swimmers?
c) What fraction of the boys do not like sports at all?

Find the fractional part of a number:

$\frac{1}{4}$ of 16 = ____ $\frac{4}{5}$ of 30 = ____ $\frac{3}{7}$ of 42 = ____ $\frac{9}{10}$ of 100 = ____

Lesson 3
Equivalent Fractions

Let us take a circle and divide it into 2 equal parts. Let's color one part yellow and the other

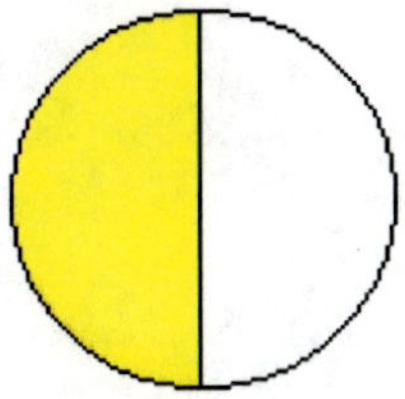

white. The fraction $\frac{1}{2}$ shows what part of the circle is yellow.

Then let us continue the division of the circle and divide each half into 2 equal parts again.

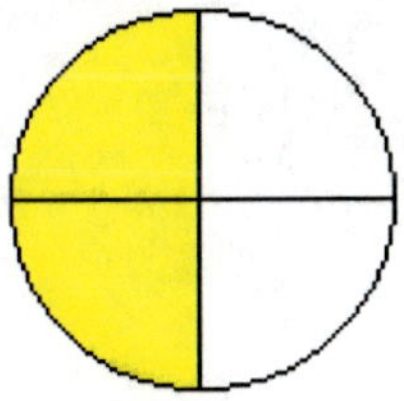

You can see 4 equal parts. The fraction $\frac{2}{4}$ shows what part of the circle is yellow.

If we continue the division of the circle and now divide each fourth into 2 equal parts, we can see that the fraction $\frac{4}{8}$ will represent the yellow part.

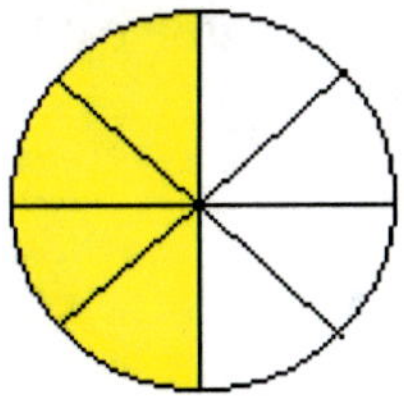

The yellow part is still the same, no matter how many times we cut it; so, we can make a conclusion that

$$\frac{1}{2}=\frac{2}{4}=\frac{4}{8}$$

Fractions that name the same amount of a region or set are called **equivalent fractions.**

Another way to show equivalent fractions is to use the number line.

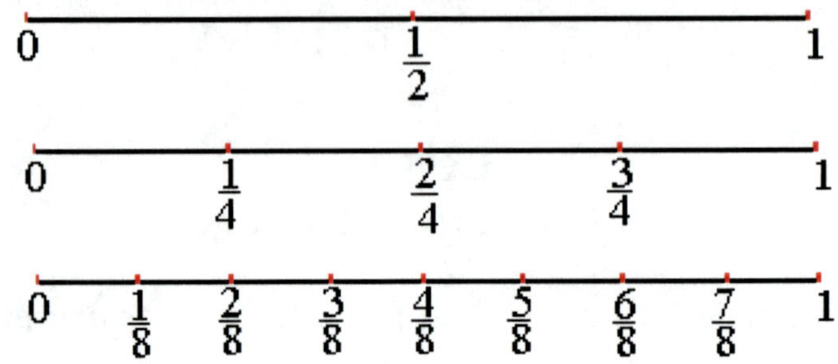

Again, you can see that $\frac{1}{2}=\frac{2}{4}=\frac{4}{8}$, and also, that $\frac{1}{4}=\frac{2}{8}$ and $\frac{3}{4}=\frac{6}{8}$.

The pictures will show you equivalent or equal fractions.

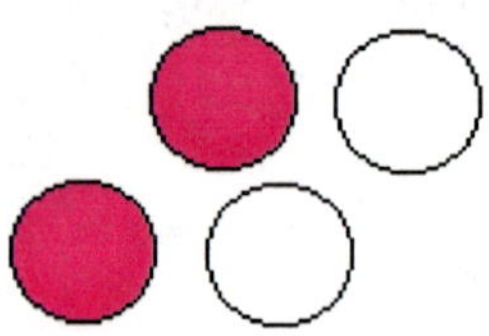

$\frac{1}{2}=\frac{2}{4}$

$\frac{1}{2}=\frac{4}{8}$

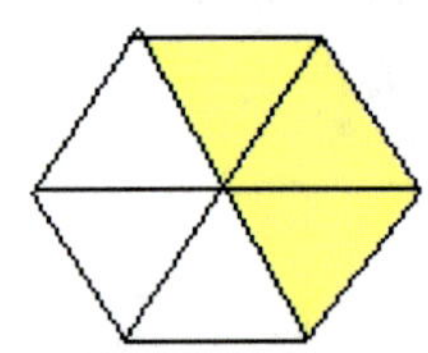

$\frac{1}{2}=\frac{3}{6}$

You have noticed that we wrote that $\frac{1}{2}=\frac{2}{4}$, then that $\frac{1}{2}=\frac{4}{8}$, then that $\frac{1}{2}=\frac{3}{6}$.
Does it mean that we can multiply the numerator and denominator by the same number and get an equivalent fraction? Yes, it does. We can also divide the numerator and denominator by the same number (if both are divisible by the number), and get an equivalent fraction.

For example: $\frac{3}{5}=\frac{3\times 4}{5\times 4}=\frac{12}{20}$ or $\frac{4}{6}=\frac{4\div 2}{6\div 2}=\frac{2}{3}$

PRACTICE

Find the equivalent fractions by multiplying the numerator and denominator by 2.

$\frac{1}{3}=$ $\frac{4}{5}=$ $\frac{5}{8}=$ $\frac{11}{12}=$ $\frac{12}{15}=$

Find the equivalent fractions by multiplying the numerator and denominator by 3.

$\frac{1}{3} =$ $\frac{4}{5} =$ $\frac{5}{8} =$ $\frac{11}{12} =$ $\frac{12}{15} =$

Give three equivalent fractions.

$\frac{2}{7} = \frac{}{} = \frac{}{} = \frac{}{}$ $\frac{4}{9} = \frac{}{} = \frac{}{} = \frac{}{}$ $\frac{5}{6} = \frac{}{} = \frac{}{} = \frac{}{}$

Lesson 4
Fractions in Lowest Terms

You already know that you can multiply or divide the numerator and denominator of a fraction by the same number and get an equal or equivalent fraction. We can always multiply each part of a fraction by the same number, but when it comes to division, things are not so easy. Sometimes we can do it, and sometimes we can't. I think that you understand that we can do it only if both the numerator and the denominator are divisible by the same number – that is, if they have a common factor.

Let us take $\frac{6}{9}$; the numerator and denominator have the common factor 3. If we divide both by 3, we will get the fraction $\frac{2}{3}$ that is equivalent to $\frac{6}{9}$.

The process of dividing the numerator and denominator by the same number is called **reducing a fraction.**

Now let us see whether or not we can reduce the fraction $\frac{2}{3}$. You will find that we can't, because the numerator and the denominator don't have any common factors - they are **relatively prime**.

A fraction is in lowest terms when the numerator and the denominator are relatively prime, or their greatest common factor is 1.

We often prefer to see fractions in lowest terms – since the numbers are small, they are easier to deal with in calculations – so let's learn how to find them.
There are two ways to find an equivalent fraction in lowest terms.
The first way is to divide both terms by common factors until the GCF equals 1.
For example, $\frac{18}{30} = \frac{18 \div 3}{30 \div 3} = \frac{6}{10} = \frac{6 \div 2}{10 \div 2} = \frac{3}{5}$

The second way is to divide both terms by their GCF.

For example, $\frac{18}{30} = \frac{18 \div 6}{30 \div 6} = \frac{3}{5}$

6 is the GCF of 18 and 30, so we divided both terms by 6 at once.

Both ways give us the same result. The second way is shorter, but sometimes it is not so easy to find the GCF for both terms. After you divide the numerator and the denominator by a common factor, the numbers will be smaller and it will be easier to find another common factor or the GCF for the new terms.
Review the material about divisibility and about the GCF (Ch.4, lesson 2 and 5), and you will be ready to learn to find fractions in the lowest terms. However, even before you review this material, you should be able to tell that if the numerator of a fraction is 1, the fraction is in lowest terms.

Now let us practice a little bit together.
We have 6 fractions. At first we will decide whether the fraction is in lowest terms or not, and if it is not, we will reduce it to lowest terms.

1) $\frac{4}{12}$ 2) $\frac{6}{8}$ 3) $\frac{7}{10}$ 4) $\frac{1}{6}$ 5) $\frac{8}{32}$ 6) $\frac{12}{18}$

1) the GCF of 4 and 12 is 4, so $\frac{4}{12} = \frac{1}{3}$

2) the GCF of 6 and 8 is 2, so $\frac{6}{8} = \frac{3}{4}$

3) the GCF of 7 and 10 is 1, so $\frac{7}{10}$ is in lowest terms

4) the numerator is 1, so the fraction $\frac{1}{6}$ is in lowest terms

5) the GCF of 8 and 32 is 8, so $\frac{8}{32} = \frac{1}{4}$

6) the GCF of 12 and 18 is 6, so $\frac{12}{18} = \frac{2}{3}$

PRACTICE

Tell whether or not the fraction is in lowest terms.

1) $\frac{3}{5}$ 2) $\frac{4}{8}$ 3) $\frac{5}{20}$ 4) $\frac{9}{12}$ 5) $\frac{10}{15}$ 6) $\frac{1}{8}$

Reduce to lowest terms when the fraction is not already in lowest terms.

7) $\frac{7}{21}$ 8) $\frac{12}{15}$ 9) $\frac{6}{11}$ 10) $\frac{28}{35}$ 11) $\frac{40}{50}$ 12) $\frac{17}{27}$

Lesson 5

Improper Fractions and Mixed Numbers

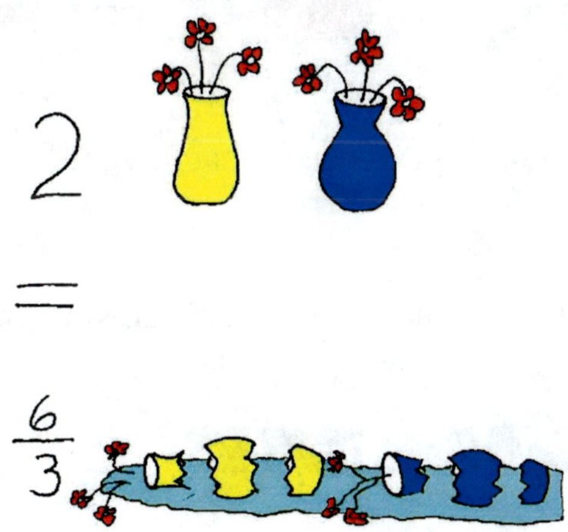

You had two apples and ate a half of one of them. What do you have left? One and a half. You can write it like this: $1\frac{1}{2}$. This is neither a whole number nor a fraction – it is a new kind of number: **a mixed number**. A whole number and a fraction make up **a mixed number**. For example, $4\frac{2}{3}$ and $9\frac{3}{7}$ are mixed numbers.

$4\frac{2}{3}$ is larger than 4 and smaller than 5. $9\frac{3}{7}$ is larger than 9 and smaller than 10.

On the number line, we can find these numbers between 4 and 5, and between 9 and 10.

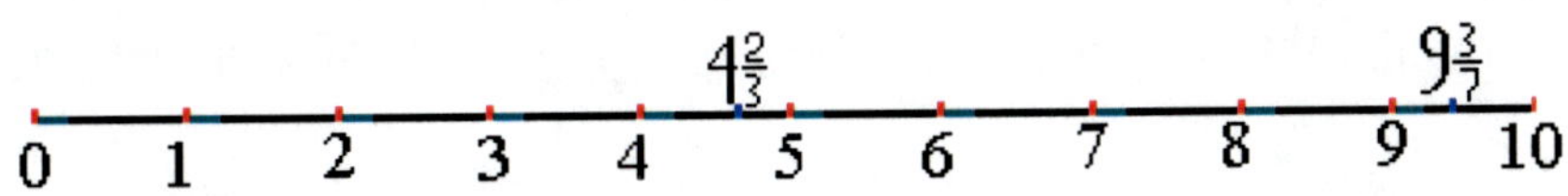

If you still have your one and a half apple, we can continue to work with them to show you a new kind of fractions. Let us divide the whole apple into 2 equal parts. How many halves do you have now? 3. We can write it like this: $\frac{3}{2}$.

For the first time, you are seeing a fraction with a numerator bigger than its denominator. Fractions with a numerator greater than or equal to the denominator are called **improper fractions**. I think you understand that these fractions are bigger than 1.

Examples of improper fractions are: $\frac{7}{3}$ and $\frac{15}{7}$.

Any mixed number can be written as an improper fraction, and any improper fraction can be written as a mixed number. Let us think about how to do it .
The pictures on the right will help us.

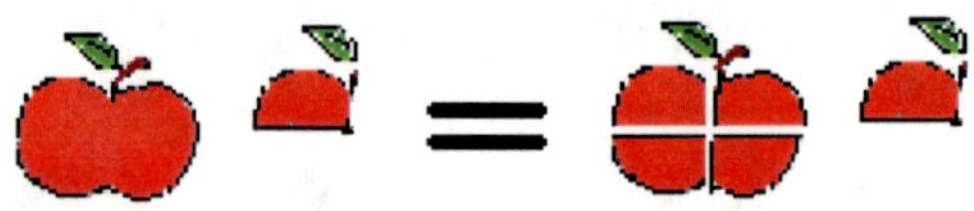

You can consider one apple as 1, or $\frac{4}{4}$. It means that you see $1\frac{1}{4}$ or $\frac{5}{4}$ of an apple.

We can consider three rectangles as 3 or $\frac{9}{3}$. It means that you see $3\frac{2}{3}$ or $\frac{11}{3}$ of a rectangle.

Here is how to change a mixed number to an improper fraction. Multiply the denominator by the whole number, add the numerator and make this sum the new numerator. The denominator will be the same.

For example, $5\frac{2}{7} = \frac{5 \times 7 + 2}{7} = \frac{37}{7}$

Here are some more examples:

$4\frac{3}{5} = \frac{4 \times 5 + 3}{5} = \frac{23}{5}$ $8\frac{4}{9} = \frac{8 \times 9 + 4}{9} = \frac{76}{9}$ $11\frac{1}{2} = \frac{11 \times 2 + 1}{2} = \frac{23}{2}$

Now is a good time to recall division with a remainder (Chapter 5, Lesson 11, Mathematics for Little Ones).
We were not lucky with each number; sometimes, we could not finish the division because the divisor was not a factor of the dividend. In this case we were talking about a remainder.
For example, $7 \div 3 = 2$ and R1 (1 is the remainder); $17 \div 5 = 3$ R2 (2 is the remainder).

After learning fractions, you can write it differently. You already know that you can divide 1 by 3 and get $\frac{1}{3}$, or divide 2 by 5 and get $\frac{2}{5}$. Therefore, we can write our quotients as

$2\frac{1}{3}$ and $3\frac{2}{5}$.

Here is how to change an improper fraction to a mixed number. Divide the numerator by the denominator; the quotient will be the whole number, the remainder will be the numerator and the denominator will be the same.

For example, $\frac{17}{3}$

```
    5 R 2
3 ) 17
   -15
     2
```

Therefore, $\frac{17}{3} = 5\frac{2}{3}$

Here are some more examples.

$\frac{19}{5} = 3\frac{4}{5}$ $\frac{29}{7} = 4\frac{1}{7}$ $\frac{30}{11} = 2\frac{8}{11}$ $\frac{100}{9} = 11\frac{1}{9}$

To gain a better understanding of this idea, you should realize that we can write any whole number as an improper fraction. For example, $2 = \frac{6}{3}$, $5 = \frac{20}{4}$, $1 = \frac{7}{7}$, $10 = \frac{100}{10}$.

Before we go ahead, I want you to know that fractions and mixed numbers are called **rational numbers.**

PRACTICE

Rewrite these mixed numbers as improper fractions.

$3\frac{4}{5} =$ $5\frac{7}{9} =$ $11\frac{5}{7} =$ $1\frac{7}{8} =$

Rewrite these improper fractions as mixed numbers in lowest terms.

$\frac{12}{7} =$ $\frac{23}{4} =$ $\frac{56}{7} =$ $\frac{100}{11} =$

Lesson 6
Word Problems

1. Bob bought 9 balloons for his birthday party. $\frac{2}{3}$ of them were red, all the others were blue. How many blue balloons did Bob buy?

2. A bag of 500 puzzle pieces has 125 green pieces. What fraction of the pieces are green? What fraction of the pieces are not green?

3. Out of 45 rows in a theater, 5 rows are for reserved seats. What fraction of the seats are reserved?

4. " I am a fraction. My numerator is 4 less than my denominator. In lowest terms I reduce to $\frac{5}{6}$. What fraction am I?"

5. Ann had 20 apples. She ate $\frac{1}{4}$ of them. How many apples are left?

6. There are 21 fruits on the plate. $\frac{2}{7}$ of them are apples, $\frac{1}{3}$ are pears, 5 bananas and the rest are kiwifruit. How many kiwifruits are on the plate?

7. A class of 27 children went to the theater. $\frac{1}{9}$ of the kids forgot their tickets at home. How many children didn't get to see the play?

8. A movie ticket for an adult costs \$ 8. A ticket for a child is only $\frac{1}{4}$ of an adult ticket. How much is a child's ticket?

9. One pound of bananas costs \$ 2. How much will $1\frac{1}{2}$ of a pound cost?

10. A class of 24 students had a math test. $\frac{2}{3}$ of them got good grades, A's and B's. $\frac{1}{6}$ of all the students failed the test. All the others got C's. How many students got C's for the test?

Chapter 5
Addition and Subtraction of Fractions

Lesson 1
Adding and Subtracting Fractions with Like Denominators

It is very easy to add or subtract fractions that have the same denominators.

Look at this apple. It is divided into 8 equal parts; therefore, we have $\frac{8}{8}$ of an apple.

What would happen if you ate $\frac{3}{8}$? You can see that you will have only $\frac{5}{8}$ of the apple left. In the language of mathematics, it looks like this:

$$\frac{8}{8}-\frac{3}{8}=\frac{5}{8}$$

Now, look at this circle.

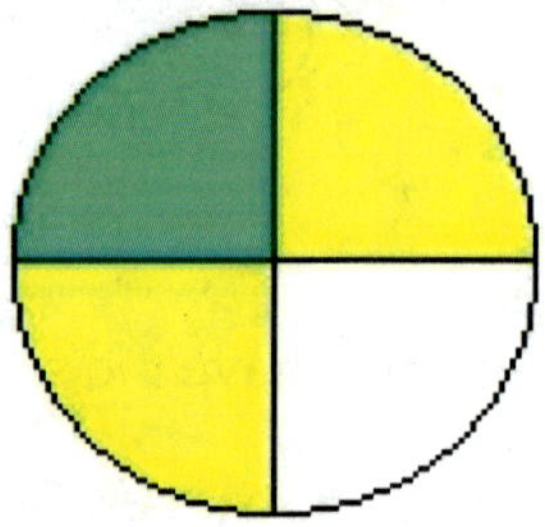

If you take the $\frac{1}{4}$ that is green and add the $\frac{2}{4}$ that are yellow, you will get $\frac{3}{4}$ that are colored.

We can write: $\frac{1}{4}+\frac{2}{4}=\frac{3}{4}$

It means that to add or subtract fractions with the same denominators (or we can say "with like denominators") we should add or subtract the numerators, and leave the denominator the same. This is easy to understand since the numerator tells us how many slices we have; if we add 1 slice to 2 slices, we will get 3 slices. However, the denominator tells us what kind of slices we have; their size doesn't change.

Let us solve several problems.

$\frac{5}{7}+\frac{4}{7}=\frac{9}{7}$ $\frac{2}{11}+\frac{3}{11}=\frac{5}{11}$ $\frac{11}{15}-\frac{2}{15}=\frac{9}{15}$ $\frac{13}{20}-\frac{3}{20}=\frac{10}{20}$

Sometimes, we can get an answer that is an improper fraction, or an answer that is a fraction that can and should be reduced. The answer to the first problem can be written as a mixed number:

$$\frac{5}{7}+\frac{4}{7}=\frac{9}{7}=1\frac{2}{7}$$

The answers to the 3rd and the 4th problems can be reduced.

$$\frac{11}{15}-\frac{2}{15}=\frac{9}{15}=\frac{3}{5} \qquad \frac{13}{20}-\frac{3}{20}=\frac{10}{20}=\frac{1}{2}$$

As with whole numbers, we can write 2 fractions horizontally, like we've been doing so far in this lesson, or vertically, writing one fraction under another.

For example:

$$\begin{array}{r}\frac{3}{11}\\ +\ \frac{5}{11}\\ \hline \frac{8}{11}\end{array} \qquad \begin{array}{r}\frac{15}{16}\\ -\ \frac{7}{16}\\ \hline \frac{8}{16}=\frac{1}{2}\end{array} \qquad \begin{array}{r}\frac{7}{8}\\ +\ \frac{5}{8}\\ \hline \frac{12}{8}=1\frac{4}{8}=1\frac{1}{2}\end{array} \qquad \begin{array}{r}\frac{21}{25}\\ -\ \frac{6}{25}\\ \hline \frac{15}{25}=\frac{3}{5}\end{array}$$

You can use any method that is easier for you.

PRACTICE

Add or subtract. Reduce if possible.

$\frac{2}{7}+\frac{3}{7}=$ $\qquad \frac{7}{15}-\frac{2}{15}=$ $\qquad \frac{1}{9}+\frac{2}{9}=$ $\qquad \frac{17}{18}-\frac{5}{18}=$

$\frac{5}{8}+\frac{1}{8}=$ $\qquad \frac{3}{10}+\frac{3}{10}=$ $\qquad \frac{5}{12}-\frac{1}{12}=$ $\qquad \frac{15}{16}-\frac{7}{16}=$

Add or subtract. Rename and reduce if possible.

$$\begin{array}{r}\frac{5}{6}\\ +\ \frac{5}{6}\\ \hline\end{array} \qquad \begin{array}{r}\frac{12}{15}\\ -\ \frac{2}{15}\\ \hline\end{array} \qquad \begin{array}{r}\frac{4}{5}\\ +\ \frac{1}{5}\\ \hline\end{array} \qquad \begin{array}{r}\frac{8}{9}\\ -\ \frac{2}{9}\\ \hline\end{array} \qquad \begin{array}{r}\frac{10}{11}\\ -\ \frac{9}{11}\\ \hline\end{array} \qquad \begin{array}{r}\frac{7}{15}\\ +\ \frac{8}{15}\\ \hline\end{array}$$

Lesson 2
Adding and Subtracting Fractions with Unlike Denominators

As you saw in Lesson 1, it is very easy to add or subtract slices that are all of the same size. However, what do you do if some of the slices are fourths and some are thirds, for example? We won't know what to call the result.

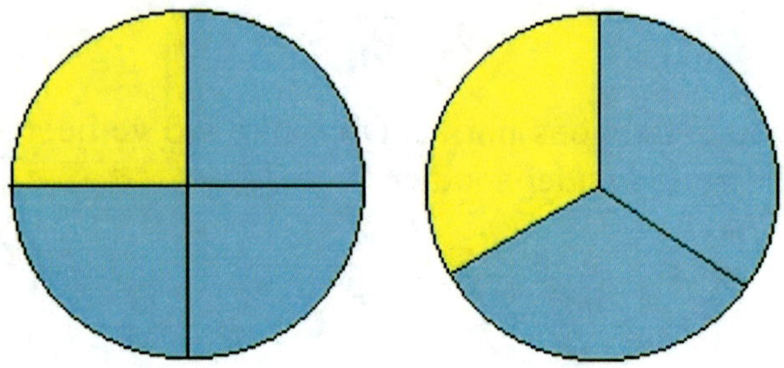

$$\frac{1}{3}+\frac{1}{4}=?$$

Remember: We Cannot Add or Subtract Fractions with Different Denominators!

So, in order to add slices of different sizes, we cut them into parts that are identical – for example, we can cut both the thirds and the fourths into twelfths, as you see in the picture.

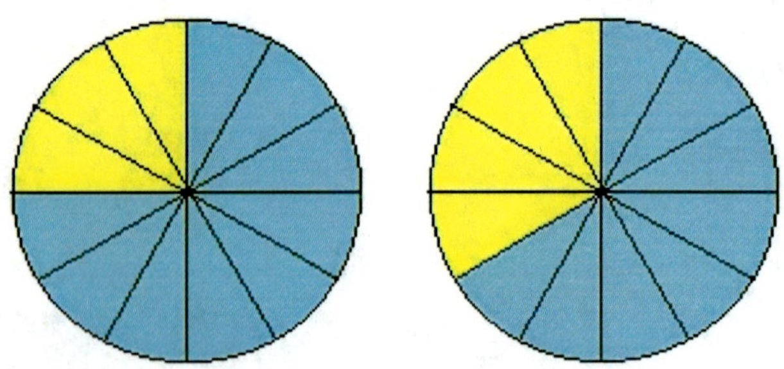

$$\frac{3}{12}+\frac{4}{12}=\frac{7}{12}$$

How did we know that we should cut them into twelfths? For the answer to this question, see the material about the Least Common Multiple in Chapter 3, Lesson 7. From what you know about equivalent fractions, you know that $\frac{1}{4}=\frac{2}{8}=\frac{3}{12}$, and $\frac{1}{3}=\frac{2}{6}=\frac{3}{9}=\frac{4}{12}$. You can see that the denominators of the equivalent fractions are all multiples of the original denominators. So, if we want to cut $\frac{1}{4}$ and $\frac{1}{3}$ into slices of the same size, we should find the number that is a multiple of 4 and also a multiple of 3, or the Least Common Multiple of 4 and 3. Then, once we have slices that are all the same size, we can add or subtract them easily.

In summary, this is what you do when you have to add or subtract fractions with different denominators:
1) Find the LCM of their denominators.
2) Write equivalent fractions with the common denominator.
3) Add or subtract.

For example, $\frac{5}{6}+\frac{3}{8}$

1) LCM (6,8) = 24

2) $\frac{5}{6}=\frac{20}{24}$ $\quad \frac{3}{8}=\frac{9}{24}$

3) $\frac{20}{24}+\frac{9}{24}=\frac{29}{24}$

Usually we write it like this:

$$\frac{5}{6}+\frac{3}{8}=\frac{20}{24}+\frac{9}{24}=\frac{29}{24}=1\frac{5}{24}$$

But again, you can write it vertically if it looks easier to you.

$$\begin{array}{l} \frac{5}{6}=\frac{20}{24} \\ + \\ \frac{3}{8}=\frac{9}{24} \\ \hline \frac{29}{24}=1\frac{5}{24} \end{array} \qquad \begin{array}{l} \frac{9}{10}=\frac{9}{10} \\ - \\ \frac{2}{5}=\frac{4}{10} \\ \hline \frac{5}{10}=\frac{1}{2} \end{array}$$

Other examples:

$$\frac{3}{4}-\frac{2}{3}=\frac{9}{12}-\frac{8}{12}=\frac{1}{12}$$

$$\frac{7}{18}+\frac{1}{6}=\frac{7}{18}+\frac{3}{18}=\frac{10}{18}=\frac{5}{9}$$

$$\frac{11}{12}+\frac{7}{8}=\frac{22}{24}+\frac{21}{24}=\frac{43}{24}=1\frac{19}{24}$$

$$\frac{9}{10}-\frac{1}{15}=\frac{27}{30}-\frac{2}{30}=\frac{25}{30}=\frac{5}{6}$$

$$\begin{array}{l} \frac{9}{11}=\frac{27}{33} \\ - \\ \frac{1}{3}=\frac{11}{33} \\ \hline \frac{16}{33} \end{array} \qquad \begin{array}{l} \frac{7}{20}=\frac{21}{60} \\ + \\ \frac{14}{15}=\frac{56}{60} \\ \hline \frac{77}{60}=1\frac{17}{60} \end{array} \qquad \begin{array}{l} \frac{7}{10}=\frac{35}{50} \\ - \\ \frac{9}{25}=\frac{18}{50} \\ \hline \frac{17}{50} \end{array}$$

PRACTICE

Add and subtract. Rename and reduce if possible.

$\frac{3}{10}+\frac{7}{8}=$ $\frac{1}{2}+\frac{7}{8}=$ $\frac{2}{5}+\frac{2}{7}=$

$\frac{6}{11}-\frac{5}{22}=$ $\frac{13}{15}-\frac{7}{12}=$ $\frac{9}{20}-\frac{1}{30}=$

$$\begin{array}{r} \frac{2}{3} \\ + \frac{4}{5} \\ \hline \end{array} \qquad \begin{array}{r} \frac{11}{12} \\ - \frac{5}{6} \\ \hline \end{array} \qquad \begin{array}{r} \frac{14}{15} \\ + \frac{4}{5} \\ \hline \end{array}$$

Lesson 3
Adding Mixed Numbers with Like Denominators

You already know how to add and subtract whole numbers and how to add and subtract fractions. The next step is addition and subtraction of mixed numbers.
At first, let us learn to add. You will do it in three steps: first add the whole parts of each number, then add the fractional parts, then add the results.
It is not too hard. Look at this example:

$$3\frac{2}{7}+5\frac{3}{7}=(3+5)+\left(\frac{2}{7}+\frac{3}{7}\right)=8+\frac{5}{7}=8\frac{5}{7}$$

Sometimes, when you add the two fractions, you can get an improper fraction. In this case you will rewrite it as a mixed number and add it to the whole number. This is exactly like addition with carrying.

For example, $7\frac{5}{8}+4\frac{7}{8}=(7+4)+\left(\frac{5}{8}+\frac{7}{8}\right)=11+\frac{12}{8}=11+1\frac{4}{8}=12\frac{1}{2}$

What happens if the first number has no fractional part? I think that you already know what to do. Just add the whole number to the whole part of the other number, and leave the fractional part alone.

For example, $12+7\frac{3}{5}=(12+7)+\frac{3}{5}=19\frac{3}{5}$

Now the vertical way of writing numbers will be even more attractive than before. Look at these examples:

$$\begin{array}{r} 7\frac{5}{8} \\ +\ 8\frac{1}{8} \\ \hline 15\frac{6}{8}=15\frac{3}{4} \end{array} \qquad \begin{array}{r} 12\frac{3}{4} \\ +\ 24\frac{3}{4} \\ \hline 36\frac{6}{4}=37\frac{2}{4}=37\frac{1}{2} \end{array} \qquad \begin{array}{r} 77\frac{14}{15} \\ +\ 23\frac{7}{15} \\ \hline 100\frac{21}{15}=101\frac{6}{15}=101\frac{2}{5} \end{array}$$

PRACTICE

Add. Reduce if possible.

$3\frac{1}{5}+4\frac{2}{5}=$ $\qquad$ $7\frac{1}{8}+5\frac{3}{8}=$ $\qquad$ $2\frac{5}{6}+9\frac{1}{6}=$

$6\frac{3}{4}+6\frac{3}{4}=$ $\qquad$ $12+11\frac{7}{10}=$ $\qquad$ $7\frac{3}{14}+8\frac{5}{14}=$

$$\begin{array}{r} 12\frac{4}{11} \\ +\ 6\frac{7}{11} \\ \hline \end{array} \qquad \begin{array}{r} 14\frac{5}{12} \\ +\ 21\frac{11}{12} \\ \hline \end{array} \qquad \begin{array}{r} 11\frac{7}{8} \\ +\ 12\frac{5}{8} \\ \hline \end{array}$$

Lesson 4
Subtracting Mixed Numbers with Like Denominators

Subtracting mixed numbers is also very easy if the first fractional part is greater than the second one. Subtract the whole parts, then subtract the fractional parts, then add the results.

For example, $8\frac{3}{5}-2\frac{1}{5}=(8-2)+\left(\frac{3}{5}-\frac{1}{5}\right)=6+\frac{2}{5}=6\frac{2}{5}$ or

$$\begin{array}{r} 17\frac{5}{6} \\ -\ 12\frac{1}{6} \\ \hline 5\frac{4}{6}=5\frac{2}{3} \end{array} \qquad \begin{array}{r} 40\frac{7}{9} \\ -\ 25\frac{4}{9} \\ \hline 15\frac{3}{9}=15\frac{1}{3} \end{array} \qquad \begin{array}{r} 22\frac{6}{11} \\ -\ 9\frac{4}{11} \\ \hline 13\frac{2}{11} \end{array}$$

However, if the first fractional part is less than the second one, things become tricky. To help you understand how to deal with this, we will start by subtracting a fraction from 1. As you know, if you have 1 pizza, it can be divided into any number of parts.

1 can be expressed as $\frac{3}{3}, \frac{4}{4}, \frac{7}{7}$, and so on.

So, if you have to subtract a fraction from 1, all you do is express 1 as a fraction with the same denominator as the fraction you are subtracting. This is easy to understand: if you have one pizza and you want to eat $\frac{1}{8}$ of it, you have to slice the pizza into 8 parts first, which can be expressed as $\frac{8}{8}$ of a pizza. Now, it is very easy to subtract $\frac{1}{8}$.

The answer is: $1-\frac{1}{8}=\frac{8}{8}-\frac{1}{8}=\frac{7}{8}$.

Now, it will be easy to subtract a fraction from any whole number. Suppose you have 5 apples and you want to eat $\frac{1}{3}$ of an apple. As you have probably guessed, what you do in this case is take 1 apple and slice it into 3 parts. As a result, instead of 5 apples, we will have $4\frac{3}{3}$. Now, it is easy to do the subtraction:

$$5-\frac{1}{3}=(4+1)-\frac{1}{3}=4+\frac{3}{3}-\frac{1}{3}=4\frac{2}{3}$$

If you understand what I have explained so far, this next step will be easy for you. If we have two mixed numbers and the first fractional part is smaller than the second one, we can use the same technique to subtract them. Let us look at this example: $5\frac{1}{4}-2\frac{3}{4}$.

We can't subtract $\frac{3}{4}$ from $\frac{1}{4}$. To solve the problem, let us take one whole apple from the five that we have, and slice it into four parts.

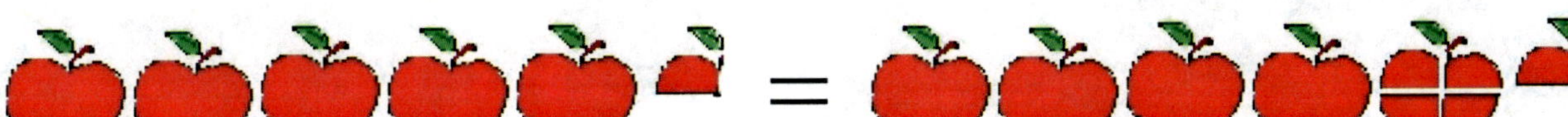

Now instead of $5\frac{1}{4}$, we have $4\frac{5}{4}$, and it is easy to find the answer. We will write it like this:

$$5\frac{1}{4}-2\frac{3}{4}=4\frac{5}{4}-2\frac{3}{4}=(4-2)+\left(\frac{5}{4}-\frac{3}{4}\right)=2\frac{2}{4}=2\frac{1}{2}$$

or this:

$$\begin{array}{r} 7\frac{1}{3} = 6\frac{4}{3} \\ -\ 2\frac{2}{3} = 2\frac{2}{3} \\ \hline 4\frac{2}{3} \end{array} \qquad \begin{array}{r} 18\frac{2}{9} = 17\frac{11}{9} \\ -\ 12\frac{5}{9} = 12\frac{5}{9} \\ \hline 5\frac{6}{9} = 5\frac{2}{3} \end{array}$$

Now let us solve several problems together.

$$1 - \frac{3}{7} = \frac{7}{7} - \frac{3}{7} = \frac{4}{7} \qquad 4 - \frac{4}{5} = 3\frac{5}{5} - \frac{4}{5} = 3\frac{1}{5}$$

$$\begin{array}{r} 10 = 9\frac{3}{3} \\ -\ 3\frac{1}{3} = 3\frac{1}{3} \\ \hline 6\frac{2}{3} \end{array} \qquad \begin{array}{r} 4\frac{1}{8} = 3\frac{9}{8} \\ -\ 1\frac{5}{8} = 1\frac{5}{8} \\ \hline 2\frac{4}{8} = 2\frac{1}{2} \end{array} \qquad \begin{array}{r} 15\frac{1}{6} = 14\frac{7}{6} \\ -\ 5\frac{5}{6} = 5\frac{5}{6} \\ \hline 9\frac{2}{6} = 9\frac{1}{3} \end{array}$$

PRACTICE

Subtract. Reduce if possible.

$12\frac{7}{8} - 8\frac{3}{8} =$ $\qquad$ $5 - \frac{3}{7} =$ $\qquad$ $7 - 4\frac{2}{5} =$

$15\frac{5}{6} - 5\frac{1}{6} =$ $\qquad$ $9\frac{3}{5} - 8\frac{4}{5} =$ $\qquad$ $18\frac{5}{9} - 5\frac{7}{9} =$

$$\begin{array}{r} 25\frac{3}{10} \\ -\ 8\frac{9}{10} \\ \hline \end{array} \qquad \begin{array}{r} 20 \\ -\ 6\frac{5}{11} \\ \hline \end{array} \qquad \begin{array}{r} 14\frac{1}{7} \\ -\ 7\frac{5}{7} \\ \hline \end{array}$$

Lesson 5

Adding and Subtracting Mixed Numbers with Unlike Denominators

Now that you know how to add or subtract mixed numbers with like denominators, it will not be hard for you to learn to do so for mixed numbers with unlike denominators. The only difference is that first we need to express the fractional parts as equivalent fractions with a common denominator. Then we can do the same thing as we did in Lessons 3 and 4.

Let us look at this example: $5\frac{1}{2}+6\frac{2}{3}$.

At first, we should find the LCD, which is 6 in this case. Then we should write equivalent fractions with the common denominator: $5\frac{3}{6}+6\frac{4}{6}$.

After this we can solve the problem easily: $5\frac{3}{6}+6\frac{4}{6}=11+\frac{7}{6}=12\frac{1}{6}$.

We write it like this: $5\frac{1}{2}+6\frac{2}{3}=5\frac{3}{6}+6\frac{4}{6}=11\frac{7}{6}=12\frac{1}{6}$

If you prefer to write numbers vertically you can do it in this case too. Again, at first you should find the common denominator, and then you will do absolutely the same thing as what you did in the previous two lessons.

Look at these examples:

$$\begin{array}{r} 12\frac{2}{3}=12\frac{6}{9} \\ -\quad 9\frac{1}{9}=9\frac{1}{9} \\ \hline 3\frac{5}{9} \end{array} \qquad \begin{array}{r} 15\frac{5}{6}=15\frac{15}{18} \\ +\quad 12\frac{7}{9}=12\frac{14}{18} \\ \hline 27\frac{29}{18}=28\frac{11}{18} \end{array} \qquad \begin{array}{r} 7\frac{1}{4}=7\frac{3}{12}=6\frac{15}{12} \\ -\quad 2\frac{5}{6}=2\frac{10}{12}=2\frac{10}{12} \\ \hline 4\frac{5}{12} \end{array}$$

Now let us solve several problems together.

$$7\frac{5}{7}+5\frac{1}{5}=7\frac{25}{35}+5\frac{7}{35}=12\frac{32}{35}$$

$$17\frac{1}{8}-11\frac{1}{3}=17\frac{3}{24}-11\frac{8}{24}=16\frac{27}{24}-11\frac{8}{24}=5\frac{19}{24}$$

$$6\frac{3}{4} = 6\frac{9}{12}$$
$$+$$
$$7\frac{2}{3} = 7\frac{8}{12}$$
$$13\frac{17}{12} = 14\frac{5}{12}$$

$$12\frac{5}{6} = 12\frac{15}{18}$$
$$+$$
$$28\frac{2}{9} = 28\frac{4}{18}$$
$$40\frac{19}{18} = 41\frac{1}{18}$$

This also works if you need to add three or more numbers; you should find the LCD of all the fractional parts, and add the whole parts and the fractional parts separately.

For example: $3\frac{1}{2}+4\frac{1}{3}+5\frac{1}{4}$

First of all we will find the LCD for 2, 3, and 4. It is 12.

Then we will write the equivalent fractions with the denominator of 12 and add these three numbers.

$$3\frac{1}{2}+4\frac{1}{3}+5\frac{1}{4} = 3\frac{6}{12}+4\frac{4}{12}+5\frac{3}{12} = 12+\frac{13}{12} = 12+1\frac{1}{12} = 13\frac{1}{12}$$

PRACTICE

Add or subtract.

$3\frac{6}{7}+5\frac{2}{3}=$ $\quad 8\frac{3}{4}-3\frac{1}{2}=$ $\quad 9\frac{7}{8}-4\frac{11}{12}=$

$14\frac{3}{10}+15\frac{4}{5}=$ $\quad 24\frac{1}{6}-10\frac{7}{8}=$ $\quad 27\frac{11}{15}+13\frac{1}{3}=$

$$100\frac{5}{9} - 38\frac{1}{6}$$

$$28\frac{3}{8} + 24\frac{6}{7}$$

$$35\frac{2}{15} - 18\frac{9}{10}$$

Lesson 6
Comparing and Ordering Fractions

Suppose you cut one chocolate bar into 2 parts, and cut another identical chocolate bar into 3 parts. Which slices will be bigger? I think that you can easily tell that the first bar will be cut into bigger pieces. In the language of mathematics, we write it like this: $\frac{1}{2} > \frac{1}{3}$. It should be obvious that the more pieces you cut something into, the smaller the pieces will be. We can say that for fractions with the same numerators, the bigger the denominator, the smaller the fraction.

Now let us divide a melon into 8 equal parts. It is obvious that $\frac{3}{8} > \frac{1}{8}$ and $\frac{3}{8} < \frac{7}{8}$. Therefore, we can say that for fractions with the same denominator, the bigger the numerator, the bigger the fraction.

Unfortunately, very often we have to compare fractions with different numerators and denominators, and this is a little bit harder. Do you remember equivalent fractions? It will help a lot. Suppose we want to compare the fractions $\frac{5}{7}$ and $\frac{8}{11}$.

It is not clear which is bigger. To make it clear, let us write equivalent fractions with a common denominator.

$$\frac{5}{7} = \frac{55}{77} \qquad\qquad \frac{8}{11} = \frac{56}{77}$$

We already know that to compare fractions with the same denominator, we have to compare the numerators. **55 < 56**, so $\frac{5}{7} < \frac{8}{11}$.

We can also use a number line to compare fractions:

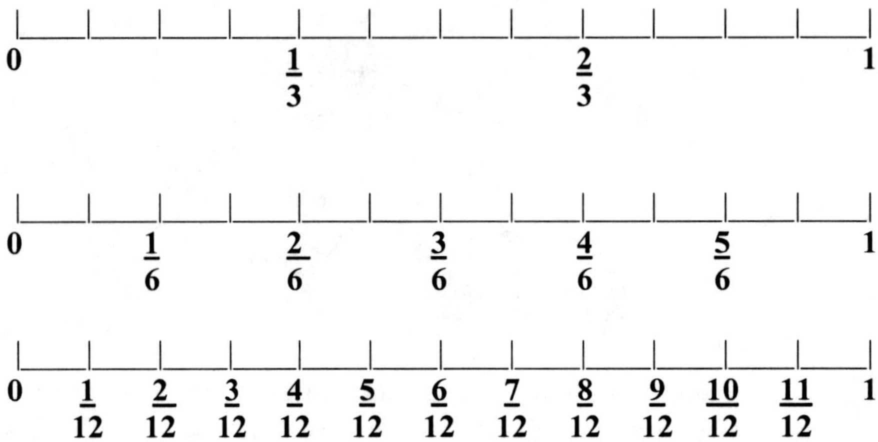

It is obvious that $\frac{1}{3} < \frac{3}{6} < \frac{9}{12}$ and $\frac{11}{12} > \frac{4}{6}$.

You can use this method to compare fractions, but it will take a lot of time to draw the number line. That is why it is better to use the first method with equivalent fractions.

For example, compare $\frac{7}{8}$ and $\frac{9}{12}$.

Write equivalent fractions with a common denominator.

$$\frac{7}{8} = \frac{21}{24} \qquad\qquad \frac{9}{12} = \frac{18}{24}$$

Now compare only the numerators. 21 is bigger than 18, therefore $\frac{7}{8} > \frac{9}{12}$.

There is another way to compare two fractions – using **cross products.** But first, let me explain to you what a cross product is. The cross product is the result of multiplying the numerator of one fraction by the denominator of another. Let us take two fractions: $\frac{2}{3}$ and $\frac{3}{5}$. 2 x 5 is the first cross product, 3 x 3 is the second cross product. Now, how can we compare two fractions? If the first cross product is greater, the first fraction is greater. Otherwise, the second fraction is greater. But if the cross products are equal, the fractions are equal too. This means that $\frac{2}{3} > \frac{3}{5}$ because **2 x 5** is greater than **3 x 3.**

Let us compare $\frac{8}{9}$ and $\frac{14}{15}$. The first cross product is **8 x 15 = 120,** the second cross product is **9 x 14 = 126. 120 < 126,** hence $\frac{8}{9} < \frac{14}{15}$.

Usually we write it like this:

$$8 \times 15 < 9 \times 14$$
$$\frac{8}{9} < \frac{14}{15}$$

Now, try to compare fractions yourself.

PRACTICE

Write >, <, or = for each □. Use equivalent fractions.

1) $\frac{3}{7}$ □ $\frac{1}{3}$ 2) $\frac{7}{8}$ □ $\frac{8}{9}$ 3) $\frac{5}{6}$ □ $\frac{2}{3}$ 4) $\frac{11}{12}$ □ $\frac{12}{15}$ 5) $\frac{6}{7}$ □ $\frac{12}{14}$

Write >, <, or = for each □. Use cross products.

6) $\frac{5}{8}$ □ $\frac{16}{25}$ 7) $\frac{11}{15}$ □ $\frac{16}{25}$ 8) $\frac{12}{14}$ □ $\frac{6}{7}$ 9) $\frac{20}{21}$ □ $\frac{11}{12}$ 10) $\frac{17}{20}$ □ $\frac{6}{7}$

Write the fractions in order from least to greatest.

11) $\frac{5}{6}, \frac{4}{5}, \frac{3}{4}$ 12) $\frac{2}{3}, \frac{5}{6}, \frac{1}{2}$ 13) $\frac{5}{8}, \frac{7}{12}, \frac{17}{24}$ 14) $\frac{8}{10}, \frac{3}{5}, \frac{11}{15}$.

Lesson 7
Estimating Sums and Differences of Fractions and Mixed Numbers

Sometimes we don't need the exact answer but only an estimate. In this case, we need not work so hard – we will use methods that will help us to find an estimate quickly that is close to the exact answer.
Here are two methods you can use to estimate the sums and differences of fractions and mixed numbers.

<u>1st method:</u> use rounding.
If a fraction is equal to or greater than a half, we increase the whole number by 1.
If a fraction is smaller than a half, we leave the whole number the same.
After this, we will work only with whole numbers, which is easy.

For example, $3\frac{1}{10} + 4\frac{5}{6}$ is about $\mathbf{3 + 5 = 8}$

Essentially, you will compare the fractional part of a number with $\frac{1}{2}$. You learned how to do it in the last lesson. Another example: $3\frac{1}{10}+4\frac{5}{6}$.

First step: $\frac{1}{10}<\frac{1}{2}$, $\frac{5}{6}>\frac{1}{2}$, therefore $3\frac{1}{10}+4\frac{5}{6}\approx 3+5=8$

2nd method: substitute compatible numbers.

Use numbers close to 0, $\frac{1}{2}$, and 1.

For example, $5\frac{9}{10}+6\frac{1}{8}+7\frac{5}{11}$

$\frac{9}{10}$ is close to 1, $\frac{1}{8}$ is close to 0, $\frac{5}{11}$ is close to $\frac{1}{2}$.

Therefore, the result is about $6+6+7\frac{1}{2}=19\frac{1}{2}$

Now look at the examples:

$5\frac{4}{5}+8\frac{1}{8}\approx 6+8=14$ (we used rounding)

$12\frac{7}{15}-6\frac{4}{9}\approx 12\frac{1}{2}-6\frac{1}{2}=6$ (we used compatible numbers)

PRACTICE

Estimate each sum or difference. Use any method you prefer.

$7\frac{1}{25}+4\frac{9}{10}\approx$

$12\frac{3}{7}-8\frac{11}{12}\approx$

$16\frac{5}{6}+13\frac{6}{7}\approx$

$19\frac{4}{9}-9\frac{5}{11}\approx$

$22\frac{7}{8}+23\frac{8}{9}\approx$

$100-29\frac{4}{13}\approx$

$50\frac{1}{8}+29\frac{2}{13}\approx$

$38\frac{1}{20}-37\frac{11}{12}\approx$

$56\frac{12}{13}-46\frac{3}{8}\approx$

Lesson 8
Word Problems

1. A triangular lot has sides of $12\frac{3}{4}$ yd, $15\frac{4}{5}$ yd, and $20\frac{7}{10}$ yd. What is the perimeter of the lot? (hint: the distance around a figure is called the perimeter of the figure) .

2. Robert read $\frac{2}{7}$ of a book the first day, and finished the book on the second day. What part of the book did Robert read on the second day?

3. Mike is $5\frac{2}{3}$ feet tall, and his sister is $4\frac{3}{7}$ feet tall. How much taller is the brother than his sister?

4. Two friends decided to meet each other. They rode their bicycles toward each other and met when one rode $2\frac{5}{6}$ miles and the other rode $2\frac{7}{8}$ miles. What is the distance between the houses?

5. A picky girl ate $\frac{1}{4}$ of an apple, $\frac{3}{4}$ of a pear, and $\frac{2}{7}$ of a banana. Which fruit did she like the least?

6. A woman made a mixture of juices: $1\frac{2}{5}$ of a liter of grape juice, $1\frac{3}{4}$ of a liter of cranberry juice and $\frac{7}{10}$ of a liter of mango juice. How many liters of mixture does she have?

7. One boy recycled $5\frac{2}{3}$ pounds of paper, another one recycled $\frac{16}{3}$ pounds. Who recycled more?

8. Mrs. Smith talked for $3\frac{3}{5}$ minutes on the phone with her sister, then $5\frac{2}{3}$ minutes with her brother, then $12\frac{7}{10}$ minutes with her daughter. How much time did she spend on the phone?

9. A baby sleeps $14\frac{1}{3}$ hours a day; a teenager sleeps $9\frac{1}{2}$ hours a day. How much more does the baby sleep than the teenager?

10. Bill ate $\frac{2}{5}$ of a watermelon one day and $\frac{3}{7}$ of the watermelon the next day. What part of the watermelon is left?

Chapter 6

Multiplication and Division of Fractions

Lesson 1

Multiplying a Fraction by a Whole Number

You remember the idea of multiplication. To multiply a number by 2 or 3 means to add the number to itself 2 or 3 times. **7 x 4 = 7 + 7 + 7 + 7.**

But what does it mean to multiply a number by a fraction? Let us take $6\times\frac{1}{3}$ as an example. We can look at this problem differently. Remember the Commutative Property of multiplication – we can change the order of the two numbers we are multiplying and get the same answer. Instead of looking at this problem as $6\times\frac{1}{3}$, let us look at it as $\frac{1}{3}\times 6$, or $\frac{1}{3}$ added to itself 6 times.

Therefore, $6\times\frac{1}{3}=\frac{1}{3}+\frac{1}{3}+\frac{1}{3}+\frac{1}{3}+\frac{1}{3}+\frac{1}{3}=\frac{6}{3}=2$

Another example: $\frac{3}{4}\times 8=\frac{3}{4}+\frac{3}{4}+\frac{3}{4}+\frac{3}{4}+\frac{3}{4}+\frac{3}{4}+\frac{3}{4}+\frac{3}{4}=\frac{24}{4}=6$

It means that **to multiply a fraction by a whole number, you should multiply the numerator by the number, and the product will be the new numerator, but the denominator will remain the same.** Then write the fraction in lowest terms if it is possible.
Let us practice a little bit.

$9\times\frac{2}{3}=\frac{18}{3}=6$ $\frac{4}{5}\times 15=\frac{60}{5}=12$ $21\times\frac{3}{7}=\frac{63}{7}=9$

There is another way of looking at this problem. What does it mean to multiply a whole number by a fraction? It means that we will take only part of the number - we should find a fractional part of the whole number. Let us take $6\times\frac{2}{3}$ as our example. You know that $\frac{2}{3}$ of something means that you divide the something into 3 parts and take 2 of them. We will do the same thing with 6 – we will divide it by 3 and multiply by 2. $6\div 3\times 2=4$

Look at the pictures below.

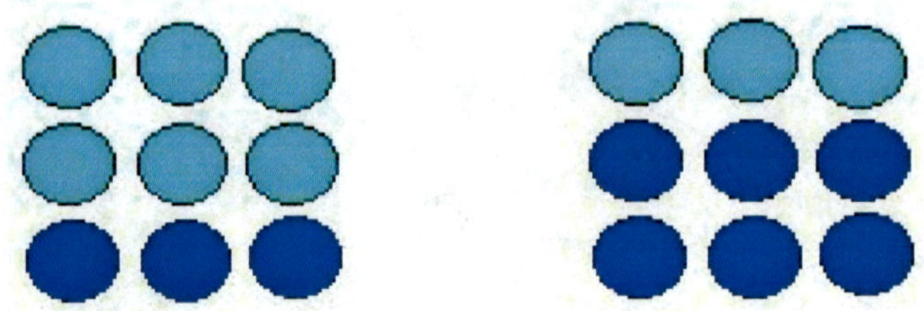

$\frac{1}{3}$ of 9 is blue $\qquad$ $\frac{2}{3}$ of 9 is blue

$\frac{1}{3}\times 9=9\div 3\times 1=3$ $\qquad$ $\frac{2}{3}\times 9=9\div 3\times 2=6$

I think you understand that **to multiply a fraction by a whole number you should divide the number by the denominator and multiply the result by the numerator of the fraction.**

For example, $\frac{1}{6}\times 12=2\times 1=2$ (we divided 12 by 6 and multiplied the result by 1)

$18\times\frac{2}{3}=6\times 2=12$ $\qquad$ $\frac{5}{7}\times 21=3\times 5=15$

You can use this method only if the whole number is divisible by the denominator of the fraction. You can use the first method in any case.

Before you start practicing by yourself, let us solve several problems together.

$\frac{3}{5}\times 25=5\times 3=15$ $\qquad$ $12\times\frac{2}{7}=\frac{24}{7}=3\frac{3}{7}$ $\qquad$ $15\times\frac{2}{3}=5\times 2=10$

$17\times\frac{3}{4}=\frac{51}{4}=12\frac{3}{4}$ $\qquad$ $\frac{7}{8}\times 16=2\times 7=14$ $\qquad$ $\frac{4}{5}\times 8=\frac{32}{5}=6\frac{2}{5}$

PRACTICE

Multiply

$18\times\frac{2}{3}=$ $\qquad$ $\frac{5}{8}\times 24=$ $\qquad$ $\frac{3}{5}\times 7=$

$\frac{5}{6}\times 36=$ $\frac{2}{7}\times 12=$ $30\times\frac{3}{10}=$

$20\times\frac{5}{9}=$ $\frac{4}{9}\times 9=$ $27\times\frac{4}{7}=$

Lesson 2
Multiplying Fractions

You already know how to find a fractional part of a whole number. Now our goal is to figure out how to find a fractional part of a fraction or a product of two fractions.

Suppose you have only $\frac{1}{2}$ of a pizza and want to eat only $\frac{1}{4}$ of this half.

Look at the picture.

You can see that it will be exactly $\frac{1}{8}$ of the pizza.

Therefore $\frac{1}{2}\times\frac{1}{4}=\frac{1}{8}$

Now, let us take a sheet of graph paper and make a 10x10 square. At first we will take only $\frac{7}{10}$ of this square and color it blue.

Then we will take only $\frac{3}{10}$ of the square and color it yellow.

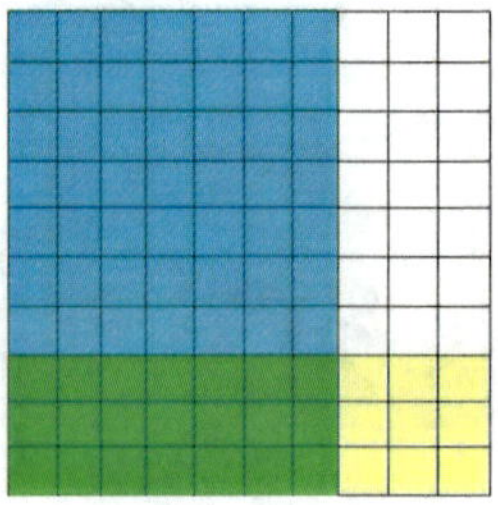

$\frac{3}{10}$ of $\frac{7}{10}$ will be green, because blue and yellow make green.

It means that $\frac{3}{10} \times \frac{7}{10} = \frac{21}{100}$ because we have 21 small green squares out of 100.

We can conclude that **to multiply two fractions we should multiply both numerators and make this product the new numerator, then multiply both denominators and make this product the new denominator.**

For example, $\frac{2}{3} \times \frac{3}{4} = \frac{6}{12} = \frac{1}{2}$

Now let us check it again with graph paper.

At first divide a rectangle into 3 equal parts with vertical lines and color only $\frac{2}{3}$ of it blue.

Then divide the rectangle into 4 equal parts with horizontal lines and color $\frac{3}{4}$ of it yellow.

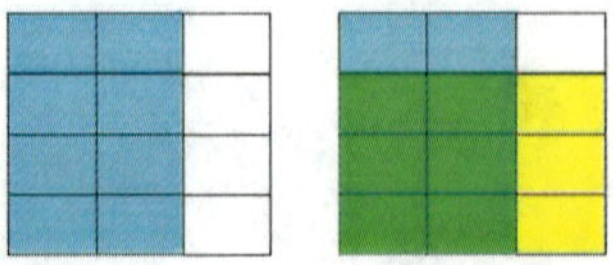

$\frac{3}{4}$ of $\frac{2}{3}$ should be green. Now count how many small squares are green. You should get only 6 out of 12, or exactly a half. Since the answer is the same, it means that the rule works.

Now look at several examples.

$\frac{3}{4}\times\frac{5}{7}=\frac{15}{28}$ $\frac{2}{5}\times\frac{7}{11}=\frac{14}{55}$ $\frac{4}{7}\times\frac{9}{10}=\frac{36}{70}=\frac{18}{35}$

You see that it is not hard to multiply fractions.

Now let us talk about reducing fractions. Of course, you can reduce the result after multiplication, as we did in the 3rd example, like this:

$$\frac{7}{8}\times\frac{4}{5}=\frac{28}{40}=\frac{7}{10} \qquad \frac{5}{12}\times\frac{3}{10}=\frac{15}{120}=\frac{1}{8}$$

However, it is much easier to reduce before multiplying. In order to reduce a fraction, you need to factor the numerator and the denominator to find common factors. It is much easier to factor a number if it is already expressed as a product of two numbers.
For example:

$$\frac{7}{8}\times\frac{4}{5}=\frac{7\times \not{4}^{1}}{\not{8}_{2}\times 5}=\frac{7}{10} \qquad \frac{5}{12}\times\frac{3}{10}=\frac{\not{5}^{1}\times \not{3}^{1}}{1\not{2}_{4}\times 1\not{0}_{2}}=\frac{1}{8}$$

In the first example, we divide both the 4 and the 8 by their common factor 4. We get 7x1 in the numerator and 2x5 in the denominator, which means that the answer is $\frac{7}{10}$. In the second example, we are luckier because we can cancel two common factors; we can divide both 3 and 12 by 3, and both 5 and 10 by 5. We get 1x1 in the numerator and 4x2 in the denominator, which means that the answer is $\frac{1}{8}$.

You see that this method allows you to work with smaller numbers, which makes calculations easier and faster.

PRACTICE

Multiply.

$\frac{3}{7}\times\frac{5}{7}=$ $\frac{4}{5}\times\frac{7}{8}=$ $\frac{2}{9}\times\frac{1}{6}=$

$\frac{8}{9}\times\frac{3}{4}=$ $\frac{6}{7}\times\frac{7}{9}=$ $\frac{1}{3}\times\frac{1}{4}=$

$\frac{7}{15}\times 5=$ $\frac{9}{16}\times\frac{12}{13}=$ $\frac{4}{51}\times\frac{17}{18}=$

Lesson 3
Multiplying Mixed Numbers

There is no special method for multiplying mixed numbers. To multiply mixed numbers, you have to write them as improper fractions. If you don't remember how to do it, review Chapter 5, Lesson 5. After you write the mixed numbers as improper fractions, you will multiply them the same way as you multiply proper fractions – multiply the numerators together and multiply the denominators together; the first product will be the new numerator, and the second one will be the new denominator. If you get an improper fraction as your answer, rewrite it as a mixed number.

For example, $3\frac{1}{4}\times 5\frac{2}{3}$

Step 1: $3\frac{1}{4}=\frac{13}{4},\quad 5\frac{2}{3}=\frac{17}{3}$

Step 2: $\frac{13}{4}\times\frac{17}{3}=\frac{221}{12}$

Step 3: $\frac{221}{12}=18\frac{5}{12}$

Here is another example, but we will write it shorter; this is the way you should write it when you do this kind of problems.

$$4\frac{1}{2}\times 5\frac{1}{3}=\frac{9}{2}\times\frac{16}{3}=\frac{144}{6}=24$$

Now, it is even more important to reduce because the numbers are larger.
In the above example, if you reduce 9 and 3 by 3 and 2 and 16 by 2 before multiplying, you will save time and avoid extra mistakes.

Look how much easier it becomes.

$$\frac{9}{2}\times\frac{16}{3}=\frac{\cancel{9}^{3}\times\cancel{16}^{8}}{\cancel{2}_{1}\times\cancel{3}_{1}}=\frac{24}{1}=24$$

Here are some more examples.

1) $1\frac{1}{2} \times 2\frac{1}{3} = \frac{\cancel{3}^1 \times 7}{2 \times \cancel{3}_1} = \frac{7}{2} = 3\frac{1}{2}$

2) $2\frac{1}{4} \times \frac{2}{3} = \frac{\cancel{9}^3 \times \cancel{2}^1}{\cancel{4}_2 \times \cancel{3}_1} = \frac{3}{2} = 1\frac{1}{2}$

3) $8\frac{1}{3} \times 2\frac{2}{5} = \frac{2\cancel{5}^5 \times 1\cancel{2}^4}{\cancel{3}_1 \times \cancel{5}_1} = \frac{20}{1} = 20$

If you have to multiply a mixed number by a whole number, rewrite the whole number as a mixed number with 1 as a denominator, and multiply the same way.

For example, $4\frac{2}{5} \times 7 = \frac{22}{5} \times \frac{7}{1} = \frac{154}{5} = 30\frac{4}{5}$

PRACTICE

Multiply

$1\frac{3}{5} \times 2\frac{3}{4} =$ $\quad 3\frac{3}{5} \times 4\frac{2}{7} =$ $\quad 7\frac{1}{7} \times 2\frac{1}{10} =$

$4\frac{7}{8} \times 8\frac{4}{9} =$ $\quad 12\frac{1}{2} \times 3\frac{1}{5} =$ $\quad 24 \times 5\frac{3}{4} =$

$\frac{8}{9} \times 6\frac{3}{4} =$ $\quad 7\frac{2}{7} \times \frac{14}{17} =$ $\quad 20\frac{1}{4} \times 3\frac{1}{9} =$

Lesson 4
Dividing Fractions

Two numbers whose product is 1 are called reciprocals of each other. What do products have to do with dividing fractions? You will understand it a little later, but for now let us learn to find reciprocals.

$\frac{2}{3}\times\frac{3}{2}=1$, so $\frac{2}{3}$ and $\frac{3}{2}$ are reciprocals of each other.

$\frac{7}{9}\times\frac{9}{7}=1$, so $\frac{7}{9}$ and $\frac{9}{7}$ are reciprocals of each other.

You can see that it is very easy to find the reciprocal of a fraction: flip the fraction over and you will have the reciprocal.

For example: $\frac{3}{5}$ and $\frac{5}{3}$, or $\frac{12}{11}$ and $\frac{11}{12}$ are reciprocals .

To find the reciprocal of a mixed number, we will do the same thing as we did to multiply mixed numbers – we will express the mixed number as an improper fraction and find its reciprocal.

For example, $3\frac{1}{2}=\frac{7}{2}$, therefore the reciprocal of $3\frac{1}{2}$ is $\frac{2}{7}$.

Now let us talk about the division of fractions.

You remember the meaning of division: we want to know how many times the divisor goes into the dividend. For example, $2\div\frac{1}{3}$ means how many thirds are in 2.

Look at the picture and you can see that there are exactly 6 thirds in 2

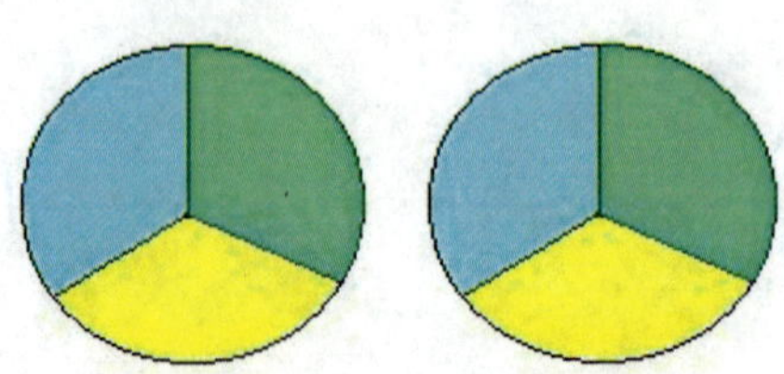

Now let us take $3\frac{1}{2}$ circles. How many $\frac{1}{4}$s are in $3\frac{1}{2}$?

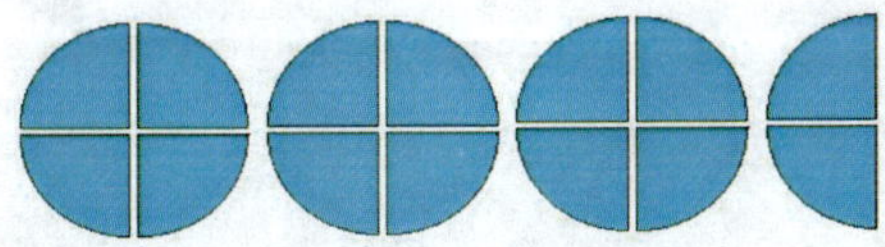

You can see that we have 14 fourths. It means that $2 \div \frac{1}{3} = 6$ and $3\frac{1}{2} \div \frac{1}{4} = 14$

You understand that the bigger the denominator, the smaller the fraction. The smaller the fraction, the more times it goes into the dividend.

Unfortunately, we can't draw pictures every time when we need to divide by a fraction. Let us look for another way. Suppose we want to divide $\frac{9}{14}$ by $\frac{5}{7}$.

Let us multiply the dividend and the divisor by $\frac{7}{5}$. We know that the quotient will be the same.

But will it help us with division? Let us see.

$$\frac{9}{14} \div \frac{5}{7} = \left(\frac{9}{14} \times \frac{7}{5}\right) \div \left(\frac{5}{7} \times \frac{7}{5}\right) = \left(\frac{9}{14} \times \frac{7}{5}\right) \div 1 = \frac{9}{14} \times \frac{7}{5}$$

Look what happened! **Instead of division by a fraction we got multiplication by the reciprocal.**

$$\frac{9}{14} \div \frac{5}{7} = \frac{9}{14} \times \frac{7}{5} = \frac{9 \times 1}{2 \times 5} = \frac{9}{10}$$

Now let us check the result using multiplication: we will multiply the quotient by the divisor and see if we get the dividend. You remember how we did it with whole numbers: $15 \div 3 = 5$ $5 \times 3 = 15$

Check: $\frac{9}{10} \times \frac{5}{7} = \frac{9 \times 1}{2 \times 7} = \frac{9}{14}$ It works !

We will solve several problems, and check the answer every time.

$\frac{2}{3} \div \frac{4}{9} = \frac{2 \times 9}{3 \times 4} = \frac{1 \times 3}{1 \times 2} = \frac{3}{2}$ Check: $\frac{3}{2} \times \frac{4}{9} = \frac{1 \times 2}{1 \times 3} = \frac{2}{3}$

$\frac{3}{5} \div 12 = \frac{3}{5} \times \frac{1}{12} = \frac{1 \times 1}{5 \times 4} = \frac{1}{20}$ Check: $\frac{1}{20} \times 12 = \frac{12}{20} = \frac{3}{5}$

$$\frac{2}{7} \div \frac{3}{28} = \frac{2}{7} \times \frac{28}{3} = \frac{2 \times 4}{1 \times 3} = 2\frac{2}{3} \qquad \text{Check: } 2\frac{2}{3} \times \frac{3}{28} = \frac{8}{3} \times \frac{3}{28} = \frac{2 \times 1}{1 \times 7} = \frac{2}{7}$$

PRACTICE

Divide.

$\frac{3}{5} \div \frac{12}{13} =$ $\qquad$ $\frac{7}{8} \div \frac{21}{22} =$ $\qquad$ $\frac{1}{11} \div \frac{8}{33} =$

$\frac{11}{12} \div \frac{11}{15} =$ $\qquad$ $\frac{8}{15} \div \frac{6}{25} =$ $\qquad$ $\frac{15}{16} \div \frac{10}{13} =$

$\frac{8}{13} \div \frac{12}{13} =$ $\qquad$ $\frac{1}{8} \div \frac{1}{10} =$ $\qquad$ $\frac{25}{27} \div \frac{100}{105} =$

Lesson 5

Dividing Mixed Numbers

It makes no difference what kind of numbers we divide. Whether we are dividing whole numbers, fractions, or mixed numbers, the idea of division is the same – we want to find out how many times the divisor goes into the dividend. It means that if we know how to divide fractions, we know how to divide mixed numbers. We should express the mixed numbers as improper fractions and then divide them the same way as we divided regular fractions. For example, we want to divide $3\frac{1}{3}$ by $6\frac{2}{3}$.

At first we will write the numbers as improper fractions, and then we will multiply the dividend by the reciprocal of the divisor.

Do you remember how to find the reciprocal of a mixed number? You learned it in the previous lesson. Write the mixed number as an improper fraction and then flip it.

Now let us solve the example, $3\frac{1}{3} \div 6\frac{2}{3}$.

$$\frac{10}{3} \div \frac{20}{3} = \frac{\cancel{10}^{1} \times \cancel{3}^{1}}{\cancel{3}_{1} \times \cancel{20}_{2}} = \frac{1 \times 1}{1 \times 2} = \frac{1}{2}$$

Don't forget to reduce before multiplying!
Let us compare 2 ways of solving the same problem: reducing before and after multiplication. Which way is easier and faster?

$$10\frac{1}{12} \div 2\frac{7}{24} = \frac{121}{12} \div \frac{55}{24} = \frac{121}{12} \times \frac{24}{55} = \frac{11 \times 2}{1 \times 5} = \frac{22}{5} = 4\frac{2}{5}$$

$$10\frac{1}{12} \div 2\frac{7}{24} = \frac{121}{12} \div \frac{55}{24} = \frac{121}{12} \times \frac{24}{55} = \frac{2904}{660} = 4\frac{264}{660} = 4\frac{2}{5}$$

In the first example you need to reduce 121 and 55 by 11, 24 and 12 by 12, and the problem becomes very easy.
In the second example you need to multiply 121 by 24 and get 2904, then multiply 12 by 55 and get 660, then divide 2904 by 660, which is not so easy. But it is not the end: then you should find the GCF of 264 and 660 and reduce the fraction.

The GCF of 264 and 660 is 132, and the fraction in lowest terms is $\frac{2}{5}$.

You see that the second way is very time-consuming, which is why I recommend you to use the first way. **Reduce everything you can**, then continue your calculations.
If we have a whole number, we should write it as an improper fraction with 1 in the denominator and do the same thing as what we did before.

Now look at several examples before you start solving problems yourself.

$$1\frac{7}{8} \div 6\frac{1}{4} = \frac{15}{8} \div \frac{25}{4} = \frac{15}{8} \times \frac{4}{25} = \frac{3 \times 1}{2 \times 5} = \frac{3}{10}$$

$$5\frac{3}{5} \div 1\frac{1}{15} = \frac{28}{5} \div \frac{16}{15} = \frac{28}{5} \times \frac{15}{16} = \frac{7 \times 3}{1 \times 4} = \frac{21}{4} = 5\frac{1}{4}$$

$$18 \div 3\frac{3}{7} = \frac{18}{1} \div \frac{24}{7} = \frac{18}{1} \times \frac{7}{24} = \frac{3 \times 7}{1 \times 4} = \frac{21}{4} = 5\frac{1}{4}$$

$$3\frac{4}{7} \div 15 = \frac{25}{7} \div \frac{15}{1} = \frac{25}{7} \times \frac{1}{15} = \frac{5 \times 1}{7 \times 3} = \frac{5}{21}$$

PRACTICE

Divide

$7\frac{1}{2} \div 2\frac{1}{2} =$ $8\frac{3}{5} \div 5\frac{1}{2} =$ $\frac{1}{3} \div 7\frac{1}{3} =$

$9\frac{3}{5} \div 2\frac{1}{10} =$ $11\frac{1}{9} \div 1\frac{7}{18} =$ $5\frac{5}{9} \div 13\frac{1}{3} =$

$1 \div 10\frac{3}{4} =$ $12\frac{1}{7} \div 5 =$ $14\frac{2}{5} \div 2\frac{6}{15} =$

Lesson 6

Estimating the Result of Multiplication or Division

Very often, getting an approximate answer very fast is more important than getting an exact answer. In this case, we will use some approximation techniques to help us do it.
The best way to estimate is to round mixed numbers to whole numbers and then work with the whole numbers, which is much easier.

For example, $6\frac{7}{8} \times 5\frac{1}{10} \approx 7 \times 5 = 35$

Let us compare this result with the exact answer.

$$6\frac{7}{8} \times 5\frac{1}{10} = \frac{55}{8} \times \frac{51}{10} = \frac{11 \times 51}{8 \times 2} = \frac{561}{16} = 35\frac{1}{16}$$

You can see that the estimate is close to the exact answer.

Here is another example.

$$12\frac{2}{9} \div 3\frac{3}{10} \approx 12 \div 3 = 4$$

Now let us find the exact answer for $12\frac{2}{9} \div 3\frac{3}{10}$.

$$\frac{110}{9} \div \frac{33}{10} = \frac{110}{9} \times \frac{10}{33} = \frac{10 \times 10}{9 \times 3} = \frac{100}{27} = 3\frac{19}{27}$$

Again you can see that the estimate is close to the exact answer.

We can also round to compatible numbers.

For example, if we want to divide $17\frac{10}{11} \div 5\frac{1}{3}$, it is better to round them to 18 ÷ 6 than to 18 ÷ 5. Of course 5 is better rounding for $5\frac{1}{3}$, but 6 is compatible with 18, and it is easier to estimate the answer.

As you go through your calculations, keep in mind that it is very easy to make a mistake. If you make one, don't worry – you are not the only one. This is why it is very important to be able to check yourself.
Before you start your calculations, think about the neighborhood of the answer.

For example, $3\frac{5}{7} \times 5\frac{7}{13}$.

Your answer can't be smaller than 3 x 5 = 15, or bigger than 4 x 6 = 24.
If your answer is within these boundaries, you can be sure that even if you are wrong, you are not very wrong.
Let us multiply.

$$3\frac{5}{7} \times 5\frac{7}{13} = \frac{26}{7} \times \frac{72}{13} = \frac{2 \times 72}{7 \times 1} = \frac{144}{7} = 20\frac{4}{7}$$

$20\frac{4}{7}$ is between 15 and 24; in the language of mathematics it looks like this:

$$15 < 20\frac{4}{7} < 24$$

Therefore our answer is correct or very close to the correct answer.

Here is another example of division: $16\frac{1}{3} \div 3\frac{1}{2}$.
I will use compatible numbers to estimate the answer. The answer should be between 16 ÷ 4 = 4 and 18 ÷ 3 = 6.
Now let us find the exact value.

$$\frac{49}{3} \div \frac{7}{2} = \frac{49}{3} \times \frac{2}{7} = \frac{7 \times 2}{3 \times 1} = \frac{14}{3} = 4\frac{2}{3}.$$

$4 < 4\frac{2}{3} < 6$. Therefore you can be sure that your answer is correct or very close to the correct answer.

PRACTICE

Estimate the product or the quotient

$6\frac{1}{7} \times 8\frac{2}{3} \approx$ $\qquad$ $12\frac{3}{10} \times 7\frac{8}{9} \approx$ $\qquad$ $14\frac{9}{10} \div 3\frac{1}{15} \approx$

$19\frac{2}{3} \div 4\frac{4}{5} \approx$ $11 \times 13\frac{3}{4} \approx$ $21\frac{4}{11} \div \frac{6}{7} \approx$

$14 \div 7\frac{1}{4} \approx$ $24\frac{8}{9} \times 3\frac{10}{13} \approx$ $35\frac{12}{13} \div 6\frac{1}{7} \approx$

Lesson 7
Word Problems

1. There are 24 students in a class. $\frac{3}{4}$ of them got A's and B's on a math test. How many students did well on the test?

2. A garden has a rectangular shape with a length of $13\frac{1}{2}$ meters and a width of $8\frac{2}{3}$ meters. Find the area of the garden. (hint: the area of a rectangle is the product of the length and width).

3. John and Larry bought a watermelon that weighs $2\frac{2}{5}$ pounds. They could eat only $\frac{5}{8}$ of the watermelon. How much was left?

4. Each class in a school lasts $\frac{3}{4}$ of an hour. How many classes do the children have if they spend 6 hours in school and have no breaks between classes?

5. Carla decided to learn Italian and read $1\frac{1}{2}$ pages of an Italian book every day. How many days will it take her to read the book if it has 33 pages?

6. How many apples are in one $5\frac{1}{4}$ pound bag, if each apple weighs $\frac{7}{8}$ of a pound?

7. A mother bought $2\frac{1}{7}$ pounds of candies and divided them fairly among her 3 kids. How many pounds of candies did each child get?

8. For a picnic, Joe, Michael, and Tom bought 16 $1\frac{1}{8}$ liter bottles of soda pop. How many liters of pop do they have?

9. A boy is 48 inches tall. He grows, on average, $1\frac{3}{4}$ inches in a year. In how many years will he be $53\frac{1}{4}$ inches tall?

10. A necklace is 20 inches long. How many beads are in this necklace if each bead is $\frac{1}{4}$ inch in diameter?

Chapter 7

Decimals: Addition and Subtraction

Lesson 1

Decimals are Fractions too

Perhaps, you already noticed that computations with fractions are much harder than computations with whole numbers. To make these computations easier, mathematicians invented a simple way of writing fractions with the denominators equal to 10, 100, 1000 and so on. They named them **decimals.**

As with whole numbers, decimals use place values. But these places will be located to the right of the ones place. Each place value is ten times the place value to its right, which is absolutely the same as in whole numbers.

At first we will separate the whole number from the fractional part with a point (we call it the **decimal point**). Then we will locate the **tenths**, which are ten times smaller than ones.

For example, we will write $\frac{3}{10}$ as 0.3, and $\frac{7}{10}$ as 0.7.

The next position to the right is the **hundredths**.

For example, we will write $\frac{7}{100}$ as 0.07, and $\frac{23}{100}$ as 0.23.

The best example of decimals is money. You know that 1 dollar equals to 10 dimes, therefore 1 dime = $\frac{1}{10}$ of a dollar, or now we can write 1 dime = 0.1 of a dollar.

5 dimes = \$0.5; 7 dimes = \$0.7 and so on.

Because one dollar equals 100 cents we can write that 1 cent = $\frac{1}{100}$ of a dollar, or 1 cent = \$ 0.01. Now, it is not difficult to see that 53 cents = \$0.53, 7 cents = \$0.07 and so on.

Money can illustrate only tenths and hundredths, but we will go further. The next position to the right of the hundredths is the **thousandths**. $\frac{1}{1000}$ = **0.001**, therefore, $\frac{9}{1000}$ =**0.009,** $\frac{57}{1000}$ = **0.057,** $\frac{277}{1000}$ = **0.277**.

The values of the first three places to the right of the decimal point are shown in the chart below. You can compare the chart to the one for whole numbers in Chapter 6, Lesson 1, Mathematics for Little Ones.

Hundreds	Tens	Ones	Tenths	Hundredths	Thousandths

Just like whole numbers, decimals can be written in expanded notation.
For example, **0.65 = 0 + 0.6 + 0.05 = 0 x (1) + 6 x (0.1) + 5 x (0. 01)**
0.279 = 0 + 0.2 + 0.07 + 0.009 = 0 x (1) + 2 x (0.1) + 7 x (0.01) + 9 x (0.001)

It is very easy to read a decimal number. You should read the entire number to the right of the decimal point as if it represented a whole number, and then give the place value of the digit farthest to the right.
For example, we will read **0.347** as: three hundred forty-seven thousandths, and **0.78** as seventy-eight hundredths.
Decimals and whole numbers may be combined to give decimals for numbers greater than 1. For example, **25.47** is read "twenty five and forty-seven hundredths", 5.693 is read "five and six hundred ninety-three thousandths".
Notice that the decimal point is read " and ".

If we have one or more zeros to the right of a decimal, we can cross them out.
It is not difficult to show that **0.300 = 0.3**
PROOF: 0.300 = 0 x (1) + 3 x (0.1) + 0 x (0.01) + 0 x (0.001)
= 0 + 3 x (0.1) + 0 + 0 = 0.3

In general, annexing zeros to the right of a decimal in this way does not change the number represented, but makes it harder to work with. So, don't write extra zeros at the end of a decimal number.

It is easy to write fractions with **10, 100, or 1000** in the denominator as decimals. But maybe you already noticed that it is also very easy to write decimals as fractions. For example, **0.3** $= \frac{3}{10}$, **0.383** $= \frac{383}{1000}$, and **7.55** $= 7\frac{55}{100}$. The only thing you should remember is that you need to reduce a fraction if it is possible. You can see that you can reduce $\frac{55}{100}$ by **5**, therefore, you can write **7.55** $= 7\frac{11}{20}$.

We will study smaller decimals in the next lesson; now practice with decimals.

PRACTICE

Write as a decimal.

1. $\frac{3}{10}$
2. $\frac{9}{100}$
3. $\frac{37}{100}$
4. $\frac{239}{1000}$
5. $5\frac{7}{10}$
6. $34\frac{23}{100}$
7. $10\frac{5}{100}$
8. $123\frac{108}{1000}$

Write in words and in expanded notation.

9. **0.6**
10. **0.04**
11. **0.008**
12. **0.029**
13. **1.2**
14. **13.26**
15. **5.056**
16. **208.207**

Write the decimals.

17. four tenths
18. twenty-five hundredths
19. three hundred sixty-six thousandths
20. seven and seven thousandths

Write as fractions.

21. **0.8**
22. **0.75**
23. **4.325**
24. **12.32**

Lesson 2
Smaller Decimals

We will follow our pattern and go further to the right of a decimal number. Now we will work with fractions with denominators of **10,000, 100,000, 1,000,000** and so on.
We will call them ten-thousandths, hundred-thousandths, millionths and so on.

Look at the chart below

Ones	Tenths	Hundredths	Thousandths	Ten-thousandths	Hundred-thousandths	Millionths
2	3	4	5	4		
5	8	5	6	4	3	
0	5	0	8	3	2	2

You can see three numbers: **2.3454, 5.85643** and **0.508322.**
They are read " two and three thousand four hundred fifty-four ten-thousandths", "five and eighty-five thousand six hundred forty-three hundred-thousandths ", and "five hundred eight thousand three hundred twenty-two millionths".

I think you agree that it is much easier to write **3.05032** than $3\frac{5032}{100000}$, although they are equal.

Now let us practice changing fractions to decimals.

$\frac{33}{100} = \mathbf{0.33}$ $\frac{257}{10000} = \mathbf{0.0257}$ $5\frac{37942}{1000000} = \mathbf{5.037942}$

You see that it is not difficult to go from fractions to decimals. Remember that the number of digits in the decimal is the same as the number of zeros in the denominator of the fraction. What is especially pleasant is that they are read the same way.

Now we will take a number written in words and write it as a decimal.
For example, we will write seventy-eight and sixty-nine ten-thousandths as **78.0069**.
Ten thousand (10,000) has four zeros, so the decimal has four digits after the decimal point.
Seven hundred forty-two thousand three hundred sixty-nine millionths is written **0.742369**.
Three and one hundred thirty-five hundred-thousandths is written **3.00135**.

To write these small decimals in expanded form will not be more difficult than what we did in the last lesson, but it will definitely be more time consuming.

0.234678 = 0 + 0.2 + 0.03 + 0.004 + 0.0006 + 0.00007 + 0.000008 =
0 x(0) + 2x(0.1) + 3x(0.01) + 4x(0.001) + 6x(0.0001) + 7x(0.00001) + 8x(0.000001)

67.06453 = 60 + 7 + 0.06 + 0.004 + 0.0005 + 0. 00003=
6 x (10) + 7 x (1) + 0 x (0.1) + 6 x (0.01) + 4 x (0.001) + 5 x (0.0001) + 3 x (0.00001)

Now let us try to write decimals as fractions.

$\mathbf{0.035} = \frac{35}{1000} = \frac{7}{200}$ $\mathbf{3.0075} = 3\frac{75}{10000} = 3\frac{3}{400}$

You see that it is easy because we pronounce equal decimals and fractions the same way:

0.73 = $\frac{73}{100}$ (seventy three hundredths),

2.271 = $2\frac{271}{1000}$ (two and two hundred seventy one thousandths), and so on.

PRACTICE

Write as a decimal

1. $\frac{57}{10000} =$
2. $\frac{12345}{100000} =$
3. $23\frac{109348}{1000000} =$
4. $\frac{7}{10000} =$
5. $6\frac{97367}{100000} =$
6. $\frac{999999}{1000000} =$

Write in words and in expanded notation.

7. **0.2323**
8. **0. 000673**
9. **4.128808**
10. **189.20002**
11. **0.000001**
12. **14.03047**

Write the decimals.

13. sixty-eight ten-thousands
14. one hundred seventy-nine thousand twenty-six millionths
15. thirty eight and forty-nine thousand one hundred sixty-two hundred-thousandths
16. seven and eight millionths

Write as fractions. Reduce if possible.

17. **0.00035**
18. **0.0125**
19. **3.0088**

Lesson 3
Comparing and Ordering Decimals

A number line can be used to show any kind of numbers. We have already used it to show whole numbers, integers, and fractions. Now, it is time to use it to show decimals.
To show tenths we will divide the interval between 0 and 1 into 10 equal parts.

0.0 0.1 0.2 0.3 0.4 0.5 0.6 0.7 0.8 0.9 1.0

We can see that as with whole numbers, a larger number is graphed to the right of a small number, so **0.7 > 0.4.**

To show hundredths, we can take any interval between two tenths and divide it again into ten equal spaces.

0.30 0.32 0.34 0.36 0.38 0.40

It is obvious that **0.32 < 0.36**.

1

3.0 3.5 A 4.0 B 4.5 5.0

Point A represents the number **3.8**, and point B represents the number half way between **4.3** and **4.4**, which is **4.35**.
Let us try to graph different points using the number line.

2.79 2.80 C 2.85 D 2.90 2.91

Point C represents the number **2.83**, and point D represents the number **2.875**.
It is very easy to compare and order decimals with a number line: the number farther to the right on the number line is the greater number.

So, **2.9 > 2.85,** **14.28 < 14.31** .

To compare decimals without a number line, start on the left, find the first place in which the digits are not equal, and compare the digits.
For example: **5.451** and **5.439**. Start on the left. The hundredths position is the first place in which the digits are not equal. Since **5 > 3, 5.451 > 5.439**.

Another example: **23.23** and **23.32**.
The tenths position is the first place in which the digits are not equal.
Since **2 < 3, 23.23 < 23.32**.
To compare decimals with a different number of decimal places, first express each number to the same number of decimal places, then compare as we did before.
For example, **0.85** and **0.841**.
You already know that **0.85 = 0.850**.
Now compare **0.850** and **0.841**. Since **5 > 4** $\Rightarrow$ **0.850 > 0.841** $\Rightarrow$ **0.85 > 0.841**.

Now, it will not be hard to order decimals.
For example, write in order from least to greatest: **3.25, 3.254, 3.251, 3.2**.
First, let us express each number to the same number of decimal places.
3.250, 3.254, 3.251, 3.200

Now you can see at once that **3.200** is the smallest number, because **0 < 5.**
Then, compare **3.250, 3.254**, and **3.251**.
Since **0 < 1 < 4, 3.250 < 3.251 < 3.254**.

The order of the numbers from least to greatest is

3.2, 3.25, 3.251, 3.254.

PRACTICE

Make a number line and draw the graph of each decimal on the number line.

1. **0.6**
2. **0.65**
3. **1.2**
4. **0.92**

Give the decimal name for points A, B, C

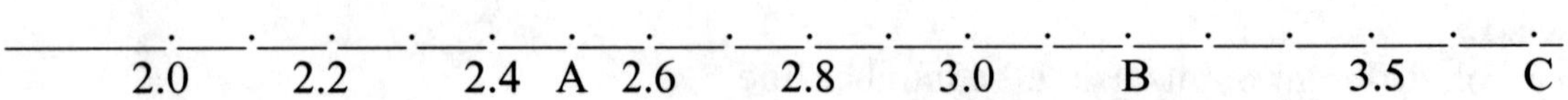

Write >, <, or = for each □.

8) **3.5** □ **3.059** 9) **0.69** □ **0.7** 10) **23.2355** □ **23.24**

Write the numbers in order from least to greatest.

11) **7.7, 7.077, 7.397**
12) **0.066, 0.072, 0.101, 0.0606**

Lesson 4
Estimating Using Rounding

This and several next lessons will show you why we prefer working with decimals to working with regular fractions. You will see that it is much easier to deal with decimals.
Let us start with rounding. We often need an estimate rather than an exact answer; this is why we will round decimals.
Look at the number line again.

0 0.1 0.5 0.8 1

Is **0.8** closer to **1** than to **0**? Yes. What about **0.1**? It is obvious that it is closer to **0** than to **1**. Therefore, we can round **0.8** to **1**, and **0.1** to **0**.

The procedure for rounding decimals is similar to the procedure for rounding whole numbers. (If you forgot it, look at Chapter 2, Lesson 2.) The difference between the two procedures is that when rounding decimals, we do not have to replace the dropped digits with zeros.

This is the procedure:
1. Underline the digit in the place to which you want to round
2. Look at the digit one place to its right:
if the digit is less than 5, round DOWN (the underlined digit will be the same)
if the digit is 5 or more, round UP (the underlined digit will increase by one)
3. Drop the digits to the right of this place (only if they are to the right of the decimal point).

Now let us round the decimals.
Round **5.7256** to the nearest tenth.
First, underline **7:** $\mathbf{5.\underline{7}256}$
Then look at the digit to the right of **7:** $\mathbf{2 < 5}$.
So, the rounded number is **5.7.**

Round **0.0382** to the nearest hundredth.
First, underline **3: 0.0382**.
Then look at the digit to the right of **3: 8 > 5**
So, the answer is **0.04**.

You can use rounding decimals to make estimates. For this you need to round each decimal, then compute with the rounded decimals. Because you still don't know how to calculate with decimals, we will round the decimals to the nearest whole numbers and then calculate.
For example, estimate the sum of **308.93 + 284.21**.
Let us round the decimals to the nearest hundred (read very carefully, we are rounding to the nearest hundred, not the nearest hundredth).
308.93 + 284.21 ≈ 300 + 300 = 600
I want you to notice that all the digits to the right of the underlined one were replaced by zeros, but only up to the decimal point. All the digits after the decimal point were dropped.

Another example: estimate the difference between **9.21** and **2.96**.
First, round each decimal to the nearest whole number, then find the difference.
9.21 - 2.96 ≈ 9 - 3 = 6
The estimate is **6**.

PRACTICE

Round to the nearest whole number.

1. **23.45** 2. **0.821** 3. **3.09** 4. **186.7**

Round to the nearest tenth.

5. **0.88** 6. **28.028** 7. **8.888** 8. **0.98**

Round to the nearest hundredth.

9. **3.2345** 10. **0.0736** 11. **45.678** 12. **0.0978**

Estimate the value by rounding.

13. **213.22 + 89.79 ≈** 14. **16.88 - 9.22 ≈**

Lesson 5
Adding and Subtracting Decimals

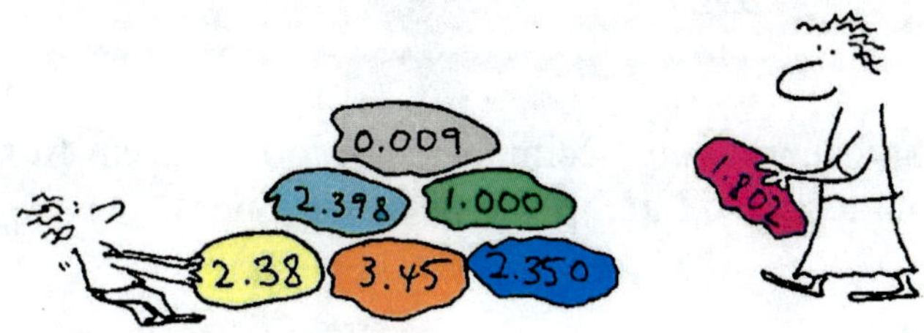

Using place value for writing decimals allows us to use the same rules for adding and subtracting decimals as we use for whole numbers.
We should understand that we can only add or subtract digits from the same decimal place, and remember that each place value to the right is 10 times smaller then its neighbor. It means that we should **line up the decimal points** when we write the given numbers one above the other. The answer will have the decimal point in line with the other decimal points.
For example, let us solve **3.52 + 5.79**.
Write the numbers one above the other and add them the same way as you add whole numbers.

$$\begin{array}{r} {\scriptstyle 1\ 1\ \ } \\ \mathbf{3.52} \\ +\underline{\mathbf{5.79}} \\ \mathbf{9.31} \end{array}$$

If the decimals have different numbers of decimal places, we can annex zeros to get them to have the same number of decimal places. You know that doing this does not change the value of the number.
For example, let us try **12.3 + 6.75**.
Since **12.3 = 12.30**, we can write

$$\begin{array}{r} {\scriptstyle 1\ \ \ \ \ \ } \\ \mathbf{12.30} \\ +\ \underline{\mathbf{6.75}} \\ \mathbf{19.05} \end{array}$$

It is even more important to annex zeros for subtraction.
For example, look at **7.5 - 0.682**.
First step: **7.5 = 7.500**
Second step:

$$\begin{array}{r} {\scriptstyle 6\ \ 14\,9\,10} \\ \mathbf{7.500} \\ -\ \underline{\mathbf{0.682}} \\ \mathbf{6.818} \end{array}$$

We can write any whole number as a decimal with an unlimited number of zeros after the decimal point, like **3 = 3.00**, or **12 = 12.000**. We will use this property when working with whole numbers and decimals at the same time.
For example, **3.52 + 7 + 0.683.**
We will write the numbers like this:

$$\begin{array}{r} {\scriptstyle 1\ 1\ \ 1\ \ \ \ \ \ } \\ \mathbf{3.520} \\ +\ \mathbf{7.000} \\ \underline{\mathbf{0.683}} \\ \mathbf{11.203} \end{array}$$

Another example: **8 - 4.68**

```
  7 9 10
  8.00
- 4.68
  3.32
```

As you can see, adding and subtracting decimals is not too different from adding and subtracting whole numbers. Just don't forget to **line up the decimal points**!

PRACTICE

Add or subtract.

1. **38.58 + 52.53**
2. **285.2 - 98.4**
3. **0.783 + 7.62**
4. **29 - 8.92**
5. **0.345 + 0.027**
6. **4.05 - 2.9**
7. **13.096 + 7**
8. **100 - 36.64**
9. **3.52 + 4.7 + 6.03** =
10. **0.23 + 0.048 + 0.672** =
11. **7.94 - 3.38** =
12. **46.2 - 8.37** =
13. Add **34.5** and **8.65**
14. Subtract **0.78** from **22**

Lesson 6
Word Problems

1. Ann had $10 and bought a chocolate bar for $5.25. How much does she have left?

2. Two friends, Michelle and Mike, went to the theater. Michelle had a reserved ticket for $6.50; Mike bought a ticket just before the performance for $13. How much did they spend for the tickets altogether?

3. John bought a math textbook for $45.75, a science textbook for $42.55, and a biology textbook for $59.99. How much did he spend for all the textbooks?

4. A boy is 1.53 meters tall. His sister is 0.27 meters shorter. How tall is the sister?

5. Two girls came to a grocery store and bought two bags of candies for $2.25 each, two ice cream cones for $1.85 each, and a can of Pepsi for $1.50. How much change did they get from a $20 bill?

6. Bill's father weighs 170.5 pounds. Together with Bill, they weigh 240.25 pounds. How much does Bill weigh?

7. A watermelon costs $2.25. A melon is $0.75 cheaper. How much should you pay to buy both?

8. Ben wants to buy two compact discs, one for $8.99 and one for $11.55. Can he afford it if he has $20?

9. Chris wants to buy a new kind of athletic shoes. If the new shoes cost $52.90 a pair, and Chris has only $25.50, how much more money does he need?

10. Sally had $13.38. Then she got $18.50 for her birthday. Is it enough to buy her favorite book if the book costs $32.50?

Chapter 8
Decimals: Multiplication and Division
Lesson 1
Multiplying Decimals

To understand how to multiply decimals, you should remember that decimals are just fractions with special denominators. A decimal's denominator can equal only **10, 100, 1000**, etc. – only 1 with one or more zeros. Let us recall how to multiply fractions. To do it, we should multiply the numerators and multiply the denominators, and then divide the first product by the second one.

It means that if we want to multiply by **0.3** ($0.3 = \frac{3}{10}$), we will multiply by **3** and divide by **10**.

For example, $20 \times 0.3 = \frac{20 \times 3}{1 \times 10} = \frac{2 \times 3}{1 \times 1} = \frac{6}{1} = 6$

Now let us think about what it means for a decimal to be divided by **10**. You know that each place value to the right is **10** times smaller. Maybe it is a good idea to move the decimal point one position to the left to make a number **10** times smaller?
Let us check this with a whole number at first. **30 ÷ 10 = 3** or we can write it this way:

30.0 ÷ 10 = 3.00

In this case it works.

Now, let us check it with a decimal. $1.5 \div 10 = 1\frac{5}{10} \div 10 = \frac{15 \times 1}{10 \times 10} = \frac{15}{100} = 0.15$

It works again!

As you see, instead of division by **10,100, 1000**, etc., we can simply move the decimal point to the left as many places as there are zeros in the divisor.
For example, **3.5 ÷ 10 = 0.35; 0.7 ÷ 100 = 0.007;**
493 ÷ 1000 = 493.0 ÷ 1000 = 0.493

I think that you understand that when we multiply by **10, 100, 1000** and so on, we should move the other way: we will move the decimal point to the right as many places as there are zeros in the factor.
For example, **3.22 x 10 = 32.2; 0.75 x 100 = 75; 63.72 x 1000 = 63,720**

Now let us talk about multiplication of decimals in general. We can multiply them like whole numbers, but then divide the product by **10** as many times as there are places to the right of the decimal point in both factors.

This is the **rule of multiplying decimals.**

1. Multiply as with whole numbers.
2. Place the decimal point in the product so that the number of places to the right of the decimal point in the product is the sum of the numbers of places to the right of the decimal point in the factors.

Examples:

2.3	1 place	**0.32**	2 places	**9.2**	1 place
x0.5	1 place	**x 0.28**	2 places	**x 15**	
1.15	2 places	**2 56**		**4 60**	
		+ 6 4		**+9 2**	
		0.0896	4 places	**138.0**	1 place

Other examples:

0.3	**3.4**	**0.75**	**24.5**
x 6	**x 2.8**	**x 0.24**	**x1.24**
1.8	**2 72**	**3 00**	**9 80**
	+ 6 8	**+ 15 0**	**+ 49 0**
	9.52	**0.18 00**	**245**
			30.380

You already know that we can drop the zeros if they are to the right of the decimal point at the end of a number. So **0.1800 = 0.18,** and **30.380 = 30.38**

To avoid mistakes in placing the decimal point, it is a good idea to estimate the answer first. For example, let us take **7.2 x 3.6.**
Mentally multiply **7 x 4**, and you will get **28.** It means that your answer should be around **28.**
Now multiply the decimals to find the exact answer.

7.2
x 3.6
19 2
+216 0
23.52

The answer is close enough to our estimate.

PRACTICE

Multiply.

1. **0.3 x 10 =** 2. **0.25 x 100 =** 3. **3.5 x 1000 =**

4. **0.7 x 0.8 =** 5. **82.2 x 0.01 =** 6. **0.62 x 0.4 =**

7. **3.02 x 0.6 =** 8. **120 x 0.28 =** 9. **21.5 x 0.32 =**

10. **17.4 x 21.3 =** 11. **35.8 x 0.38 =** 12. **0.03 x 0.573 =**

Lesson 2
Dividing Decimals

Before you start dividing decimals, I recommend you to review division of whole numbers. (Chapter 1, Lessons 4, 5, 8).
Let us start with the easiest kind of division – dividing decimals by **10, 100, 1000**, etc. I think you understand that when you divide a number by **10**, for example, each position of the number becomes **10** times smaller – tens become ones, ones become tenths, tenths become hundredths and so on. It means that all we have to do is move the decimal point one position to the left, like this: **25.37 ÷ 10 = 2.537**. It makes sense that to divide a decimal by **100, 1000**, etc., all you do is move the decimal point to the left as many places as there are zeros in the divisor. For example, **3.68 ÷ 100 = 0.0368; 67.03 ÷ 1000 = 0.06703**
If division of whole numbers is not a problem for you, division of a decimal by a whole number will not be a problem either. The idea of division is the same: separating a number into several equal parts. Do you see a big difference between dividing **6** by **3** and dividing **0.6** by **3**? No. In the first case the answer is **2**, in the second – **2** tenths. We can check the answer by multiplying: **3 x 2 = 6, 3 x 0.2 = 0.6.**
This was too easy. Let us try something harder. For example, **1.2 ÷ 4.** To do this, just remember that one is ten tenths.

1.2 ÷ 4 = 0.3 (check: **0.3 x 4 = 1.2**)

Remember that each place value to the left is **10** times bigger than the next place value to the right in a whole number as well as in a decimal, so, to divide a decimal by a whole number, we will proceed as if we were dividing two whole numbers. The only difference is the placement of the decimal point in the quotient. Common sense tells us that **the decimal point should be directly over the decimal point in the dividend.**

For example,

```
     2.2
7 ) 15.4
    -14
      1 4
      1 4
        0
```

Check:

```
  2.2
x   7
 15.4
```

Other examples.

```
    0.047
6 ) 0.282
     24
      42
      42
       0
```

```
    0.00378
9 ) 0.03402
       27
        70
        63
         72
         72
          0
```

```
      2.4
18 ) 43.2
     36
      7 2
      7 2
        0
```

Sometimes a division does not terminate, or does not come out evenly. In this case we can annex zeros to the end of the dividend, which, as you already know, does not change the value of the decimal.

For example,

```
    0.64
5 ) 3.24
    3 0
      24
      20
       4
```

we can rewrite

```
    0.648
5 ) 3.240
    3 0
      24
      20
       40
       40
        0
```

We can annex as many zeros as we wish.

More examples.

```
    0.1125
4 ) 0.4500
      4
       5
       4
       10
        8
        20
        20
         0
```

```
    0.0125
8 ) 0.1000
      8
      20
      16
       40
       40
        0
```

```
      0.24
25 ) 6.00
     5 0
     1 00
     1 00
        0
```

Check: **0.1125**	Check: **0.0125**	Check: **0.24**
x 4	**x 8**	**x 25**
0. 4500 = 0.45	**0.1000 = 0.1**	**1 20**
		4 8
		6. 00 = 6

PRACTICE

Divide and check.

1. **3.75 ÷ 10 =**
2. **0.841 ÷ 100 =**
3. **26.43 ÷ 1,000 =**
4. **0.35 ÷ 7 =**
5. **4.11 ÷ 3 =**
6. **17.022 ÷ 6 =**
7. **34.75 ÷ 2 =**
8. **0.1116 ÷ 9 =**
9. **0.084 ÷ 12 =**

Lesson 3
Dividing Decimals (Repeating Decimals)

So far, we have been lucky – we were able to divide numbers evenly without a remainder. The kind of decimals we got as answers is called **terminated decimals.** Unfortunately, we are not always so lucky. Look at this example: **0.4 ÷ 3.**

```
     0.1333...
3 )  0.4000...
      3
      10
       9
       10
        9
        10   (this keeps repeating)
         9
```

This kind of decimals is called **repeating decimals** and to write them, we will use special notation – **bar notation** – because, obviously, we can't keep writing these digits forever. To show that one or more digits keep repeating forever, we will write a bar above this digit or digits.

For example, we will write our answer **0.1333...** as $\mathbf{0.1\overline{3}}$.

Other examples: **0.343434... =** $\mathbf{0.\overline{34}}$,   **7.023232...=** $\mathbf{7.0\overline{23}}$

Now you can write each fraction or mixed number as a decimal, terminated or repeating.
To write a fraction as a decimal, divide the numerator by the denominator.
Let us try it.

```
         0. 66...
2/3 = 3 ) 2. 00
          1 8
           20
           18
            2
```

```
          0.466..
7/15 = 15 ) 7.00
            6 0
            1 00
              90
              10
```

```
                    12.75
12 3/4 = 51/4 = 4 ) 51.00
                    4
                    11
                     8
                     30
                     28
                      20
                      20
                       0
```

So, $\frac{2}{3}=0.\overline{6}$ $\frac{7}{15}=0.4\overline{6}$ $12\frac{3}{4}=12.75$

To write a mixed number as a decimal, it is not necessary to rewrite it as an improper fraction first. You can change only the fractional part, but the whole part will stay the same.

This is confirmed by this example: $12\frac{3}{4}=12.75$

More examples: $8\frac{2}{3}$ or $7\frac{7}{15}$ (Use the answers from the examples above)

$8\frac{2}{3}=8.\overline{6}$ and $7\frac{7}{15}=7.4\overline{6}$

These examples of repeating decimals were not difficult, but sometimes we have to do long, long calculations before we find the repeating part.

For example, **3 ÷ 7** or **15 ÷ 13.** Look!

$$
\begin{array}{r}
0.4285714\ldots \\
7\ \overline{)\ 3.0000000} \\
\underline{2\,8} \\
20 \\
\underline{14} \\
60 \\
\underline{56} \\
40 \\
\underline{35} \\
50 \\
\underline{49} \\
10 \\
\underline{7} \\
30 \\
\underline{28} \\
2
\end{array}
\qquad\qquad
\begin{array}{r}
1.1538461\ldots \\
13\ \overline{)\ 15.0000000} \\
\underline{13} \\
2\,0 \\
\underline{1\,3} \\
70 \\
\underline{65} \\
50 \\
\underline{39} \\
110 \\
\underline{104} \\
60 \\
\underline{52} \\
80 \\
\underline{78} \\
2
\end{array}
$$

It means that $\frac{3}{7} = 0.\overline{428571}$ and $\frac{15}{13} = 1.\overline{153846}$

You see that it is very time-consuming, but a lot of fun!
Now try some problems yourself and compare the answers: $\mathbf{5 \div 11}$; $\mathbf{41 \div 333}$.

$$
\begin{array}{r}
0.45\ldots \\
11\ \overline{)\ 5.00} \\
\underline{4\,4} \\
60 \\
\underline{55} \\
5
\end{array}
\qquad\qquad
\begin{array}{r}
0.123\ldots \\
333\ \overline{)\ 41.000} \\
\underline{33\,3} \\
7\,70 \\
\underline{6\,66} \\
1\,040 \\
\underline{999} \\
41
\end{array}
$$

So, $\frac{5}{11} = 0.\overline{45}$, $\frac{41}{333} = 0.\overline{123}$

Now that you know how to express any fraction as a decimal, you will find that it is much easier to compare fractions if you write them as decimals first. For example, which fraction is bigger: $\frac{7}{11}$ or $\frac{8}{13}$?

Let us express them as decimals and find out.

$$7 \div 11 = \begin{array}{r} 0.63\ldots \\ 11\,\overline{)\,7.00} \\ \underline{66} \\ 40 \\ \underline{33} \\ 7 \end{array} \qquad\qquad 8 \div 13 = \begin{array}{r} 0.61\ldots \\ 13\,\overline{)\,8.00} \\ \underline{78} \\ 20 \\ \underline{13} \\ 7 \end{array}$$

0.63… is greater than **0.61…**, hence $\frac{7}{11} > \frac{8}{13}$.

PRACTICE

Write each repeating decimal using bar notation.

1. **0.272727... =** 2. **3.245555... =** 3. **0.456456456... =**

Divide.

5. **0.5 ÷ 3 =** 6. **1.2 ÷ 9 =** 7. **0.23 ÷ 10 =**

8. **0.47 ÷ 7 =** 9. **3.2 ÷ 100 =** 9. **37.4 ÷ 6 =**

Change each fraction to a decimal.

10. $\frac{2}{11} =$ 11. $\frac{35}{8} =$ 12. $9\frac{4}{7} =$

Lesson 4
Approximation of a Decimal Quotient

When a division does not terminate, we may wish to find an approximate quotient to a specified number of decimal places. To do this, we should divide to one place beyond this number. After division, we will round the quotient to the specified number of decimal places.
It is possible that we will have to annex zeros to the end of the dividend, but we already know that it does not change the value of the decimal.
For example, let us divide **3.45** by **7** to the nearest hundredth.
To round to the nearest hundredth, we should know what digit is in the thousandths place, which is why we will annex one zero to the end of the dividend . So, we will divide **3.450** by **7**.

```
     0.492 ..
 7 ) 3.450
     28
      65
      63
       20
       14
        6
```

Now we can round **0.492** to the nearest hundredth.

0.492 ≈ 0.49

This technique is also useful for problems that have nothing to do with repeating decimals. We often know in advance that we don't need a precise answer, only an approximation. In this case, if we know how many decimal places we need in the answer, we will divide to one place beyond that, and then round to the required place.

For example, divide **25** by **12** to the nearest tenth.
First annex two zeros to the end of the dividend.

```
       2.08
 12 ) 25.00
      24
       100
        96
         4
```

Rounding **2.08** to the nearest tenth gives **2.1**

Another example: divide **0.35** by **8** to the nearest thousandth.
Annex two zeros to the end of **0.35**.
Then divide to the ten-thousandth.

```
     0.0437
 8 ) 0.3500
       32
        30
        24
         60
         56
          4
```

Rounding **0.0437** to the nearest thousandth gives **0.044**.

PRACTICE

Divide. Round to the nearest tenth.

1. **11 ÷ 6 =** 2. **3.2 ÷ 7 =** 3. **1.7 ÷ 8 =**

Divide. Round to the nearest hundredth.

4. **4.52 ÷ 9 =** 5. **15 ÷ 13 =** 6. **0.03 ÷ 4 =**

Divide. Round to the nearest thousandth.

7. **16 ÷ 7 =** 8. **0.53 ÷ 20 =** 9. **11 ÷ 12 =**

Lesson 5
Dividing by a Decimal

You already know that if we multiply or divide the dividend and the divisor by the same number, the quotient will be the same. For example: **600 ÷ 300 = 2, 60 ÷ 30 = 2,**
6 ÷ 3 = 2.
Does it mean that **0.6 ÷ 0.3 = 2**? Let us check it using multiplication. **2 x 0.3 = 0.6.**
It works! That means that we know how to divide by a decimal. To do this, we will multiply the dividend and the divisor by **10** as many times as we need to make the divisor a whole number. After this our task will be the same as before – to divide decimals or whole numbers by whole numbers. We already know how to do this.
For example, **0.12 ÷ 0.4.**
To make the divisor a whole number, we need to multiply it by **10**. So we will multiply both the dividend and the divisor by 10. We will get **1.2 ÷ 4.**
You already know how to solve this kind of problems; the answer is **0.3.**

You remember that to multiply a decimal by **10, 100, 1000** and so on we should move the decimal point to the right as many places as there are zeros in the factor (Ch.1, Lesson 1). Because of this, our problem can be solved easily. Look at the divisor first and move the decimal point to the right as many places as you need to make it a whole number. Then move the decimal point of the dividend the same number of places. After these two steps, divide.

For example, **0.00234 ÷ 0.003.**

$$0.00234 \div 0.003 = 2.34 \div 3$$

Now we have: **2.34 ÷ 3 = 0.78**

Let us practice a little bit doing only the first step, making the divisor a whole number:

0.082 ÷ 4.1 = 0.82 ÷ 41
3.51 ÷ 0.003 = 3510 ÷ 3
0.08722 ÷ 0.89 = 8.722 ÷ 89

After the first step, you have a problem that you know how to solve: division of decimals by whole numbers. This is exactly what you did in the last lesson.

Now let us solve these problems doing both steps and finding the quotients.
To be sure that your answer is correct, don't forget to check the result by multiplication. Multiply the quotient by the divisor; if the product is the dividend, you solved the problem correctly. If the product is different from the dividend, you made a mistake. In this case, do the problem again and be careful this time.

0.082 ÷ 4.1 = 0.82 ÷ 41 = 0.02

Check:
$$\begin{array}{r} 4.1 \\ \times\ \underline{0.02} \\ 0.082 \end{array}$$

3.51 ÷ 0.003 = 3510 ÷ 3 = 1170

Check:
$$\begin{array}{r} 1170 \\ \times\ \underline{0.003} \\ 3.510 \end{array}$$

0.08722 ÷ 0.89 = 8.722 ÷ 89 = 0.098

Check:
$$\begin{array}{r} 0.098 \\ \times\ \underline{\ \ 0.89} \\ 882 \\ +\ \underline{\ \ 7840} \\ 0.08722 \end{array}$$

More examples.

4.9 ÷ 0.007 = 4900 ÷ 7 = 700

0.00387 ÷ 0.09 = 0.387 ÷ 9 = 0.043

$$3.7\ \overline{)\ 24.05} \quad = \quad \begin{array}{r} 6.5 \\ 37\ \overline{)\ 240.5} \\ \underline{222}\ \ \ \\ 18\ 5 \\ \underline{18\ 5} \\ 0 \end{array}$$

$$0.012\ \overline{)\ 0.2244} \quad = \quad \begin{array}{r} 18.7 \\ 12\ \overline{)\ 224.4} \\ \underline{12}\ \ \ \ \ \\ 104\ \ \ \\ \underline{96}\ \ \ \\ 8\ 4 \\ \underline{8\ 4} \\ 0 \end{array}$$

Just as with division by whole numbers, we can round the answer to a specified number of decimal places.
For example, divide **0.037** by **0.06** and round to the nearest hundredth. I think you understand that you should divide until you get thousandths. In this case, you can round to the nearest hundredth.

$$0.037 \div 0.06 = 3.7 \div 6 = \begin{array}{r} 0.616 \approx 0.62 \\ 6\ \overline{)\ 3.700} \\ \underline{36}\ \ \ \ \\ 10\ \ \ \\ \underline{6}\ \ \ \\ 40\ \ \\ \underline{36}\ \ \\ 4\ \ \end{array}$$

Another example. Divide **48.2** by **0.7** and round to the nearest tenth. Now you should divide until you get hundredths.

$$48.2 \div 0.7 = 482 \div 7 = 7\overline{)482.00} = 68.85 \approx 68.9$$

```
        68.85 ≈68.9
    7 ) 482.00
        42
         62
         56
          60
          56
           40
           35
            5
```

PRACTICE

Divide

1. **0.3 ÷ 0.06 =** 2. **2.8 ÷ 0.7 =** 3. **1.21 ÷ 0.011 =**

4. **40.29 ÷ 0.51 =** 5. **10 ÷ 0.3 =** 6. **2.916 ÷ 5.4 =**

Divide and round to the nearest hundredth

7. **0.07 ÷ 0.3 =** 8. **3.58 ÷ 0.7 =** 9. **0.12 ÷ 6.7 =**

Lesson 6
Word Problems

1. A rectangular garden is 12.5 meters long and 8.3 meters wide. Find the area of the garden.

2. The area of a rectangular garden is 145.08 square feet. What is the length of the garden if the width is 7.8 feet?

3. Susan works 6 hours per day at $5.25 per hour. How much does she earn in 7 days?

4. Jones' store has a price of 5 for $42 on video games. Smith's store is selling 8 video games for $63. Which store has the better unit price for video games?

5. Jake has 15 minutes left to finish a test, and 7 problems left to solve. Will he be able to solve all the problems if it takes him 2.5 minutes to solve each one?

6. Jenny likes candies. She had $2.5 and spent it all on candies that cost $0.25 each. How many candies did she buy?

7. The distance between two cities is 227.5 miles. A family traveled this distance in 3.5 hours. What was their speed?

8. Three boys decided to buy a ping-pong table for $96.36. How much will each boy pay if they split the expenses equally?

9. Tamara bought 2.8 pounds of apples. How much did she pay, if a pound of apples costs $0.75?

10. A frog and a toad swim toward each other from opposite sides of a pond. The frog can swim at 7.8 inches per second, and the toad at 8.6 inches per second. If they meet in 6.5 second, how wide is the pond?

Chapter 9

Combined Operations with Different Kind of Numbers

Lesson 1

Combined Operations with Fractions and Mixed Numbers

If you know all four operations with fractions and mixed numbers, combining these operations will not be a problem. Fortunately, the order of operations for these numbers is the same as for whole numbers. Let us recall it.

ORDER OF OPERATIONS

1. Do all the operations inside the parentheses
2. Do all multiplication and division in order from left to right
3. Do all addition and subtraction in order from left to right

For example, look at this: $\frac{1}{2}+\frac{1}{2}\times\frac{2}{5}$

At first we will multiply, $\frac{1}{2}\times\frac{2}{5}=\frac{1}{5}$. The next step is addition: $\frac{1}{2}+\frac{1}{5}=\frac{5}{10}+\frac{2}{10}=\frac{7}{10}$

The same goes for mixed numbers.

For example, $2\frac{1}{2}-3\frac{1}{3}\div 6\frac{2}{3}$.

The first step is division: $3\frac{1}{3}\div 6\frac{2}{3}=\frac{10}{3}\div\frac{20}{3}=\frac{10\times 3}{3\times 20}=\frac{1}{2}$

The next step is subtraction: $2\frac{1}{2}-\frac{1}{2}=2$

You know that every whole number can be written as an improper fraction, so you can solve problems with mixed and whole numbers easily.

For example, $3\times\left(7\frac{3}{4}-2\frac{2}{3}\right)-2\times\frac{5}{6}$.

First step: $7\frac{3}{4}-2\frac{2}{3}=5\frac{9}{12}-\frac{8}{12}=5\frac{1}{12}$

Second step: $3\times5\frac{1}{12}=\frac{3}{1}\times\frac{61}{12}=\frac{61}{4}=15\frac{1}{4}$

Third step: $2\times\frac{5}{6}=\frac{2\times5}{1\times6}=\frac{5}{3}=1\frac{2}{3}$

Fourth step: $15\frac{1}{4}-1\frac{2}{3}=15\frac{3}{12}-1\frac{8}{12}=14\frac{15}{12}-1\frac{8}{12}=13\frac{7}{12}$

So, the rules are the same, but solving the problems is much more time-consuming. You need to be very careful to avoid mistakes, and you should practice a lot to be good at this kind of problems. Look at several more examples and then practice.

More examples.

$$17-2\frac{1}{2}\times3\frac{1}{5}=17-\frac{5\times16}{2\times5}=17-8=9$$

$$15\frac{2}{5}+4\times\left(3\frac{1}{5}-1\frac{3}{20}\right)=15\frac{2}{5}+4\times\left(3\frac{4}{20}-1\frac{3}{20}\right)=15\frac{2}{5}+4\times2\frac{1}{20}=15\frac{2}{5}+8\frac{1}{5}=23\frac{3}{5}$$

$$\frac{5}{8}\div\frac{15}{16}-\frac{3}{4}\times\frac{2}{3}=\frac{5\times16}{8\times15}-\frac{3\times2}{4\times3}=\frac{2}{3}-\frac{1}{2}=\frac{4}{6}-\frac{3}{6}=\frac{1}{6}$$

PRACTICE

$\frac{3}{8}\times\frac{4}{9}+\frac{2}{3}\times\frac{3}{4}=$

$32-3\frac{1}{2}\times6\frac{2}{7}=$

$8\frac{3}{4}\times2\frac{2}{5}-3\times\frac{4}{9}=$

$7\times\left[3\frac{1}{2}-2\times\left(3\frac{1}{2}-2\frac{1}{4}\right)\right]=$

$6\frac{2}{5}-3\frac{3}{8}\div 11\frac{1}{4}=$

$12\frac{1}{4}\div\frac{7}{16}-3\div\frac{1}{6}=$

$8\frac{1}{3}+3\times\left(5-1\frac{1}{9}\right)=$

$7\frac{1}{2}+3\frac{3}{4}-3\frac{1}{3}\times 2\frac{1}{4}=$

Lesson 2
Combined Operations with Decimals

You already know that decimals are fractions too, but fractions with special denominators. You will not be surprised that the order of operations with decimals is the same as with fractions. If you forgot how to do all four operations with decimals, review Chapter 7, Lesson 5 and Chapter 8, Lessons 1, 2, 3, and 5 and you will be ready to solve problems with combined operations with decimals.
For example, let's take **0.3 + 0.4 x 0.5**.
First, we will multiply **0.4** by **0.5**, then add **0.3** to the result of multiplication.

0.3 + 0.4 x 0. 5 = 0.3 + 0.2 = 0.5

Another example: **(0.3 + 0.4) x 0.5**.
Now we will do the addition first, because it is in parentheses. Then we will multiply.
(0.3 + 0.4) x 0.5 = 0.7 x 0.5 = 0.35
Working with decimals and whole numbers doesn't change anything in the order of operations.
For example, **5 x 3.7 + 4 ÷ 0.5 = 18.5 + 8 = 26.5**.
At first we did the multiplication, then the division, and only then the addition.

Remember that we can show division 3 ways: **6 ÷ 3 = $3\overline{)6}$ = $\frac{6}{3}$**

For combined operations, we use the third way very often.
For example, instead of writing
(3.2 + 4.8) ÷ (4.4 - 0.6 x 4), we can write $\dfrac{\mathbf{3.2 + 4.8}}{\mathbf{4.4 - 0.6 \times 4}}$.

For calculations, however, we should keep in mind that the order of operations will be the same, i.e. first calculate the numerator and the denominator separately, and only then divide the numerator by the denominator.

For example, $\dfrac{3.2+4.8}{4.4-0.6\times 4}=\dfrac{8}{4.4-2.4}=\dfrac{8}{2}=4$

More examples:

$$\frac{2.2+2\times 0.4}{0.6\div 3+0.3}=\frac{2.2+0.8}{0.2+0.3}=\frac{3}{0.5}=6$$

$$\frac{0.4\times 0.3-0.2\times 0.1}{0.8\div 0.2}=\frac{0.12-0.02}{4}=\frac{0.1}{4}=0.025$$

$$1.4-\frac{0.8\times 3.5}{3.4-1.4}=1.4-\frac{2.8}{2}=1.4-1.4=0$$

PRACTICE

1) **0.3 x 0.5 - 0.2 x 0.6 =**

2) **3.28 - 6.4 ÷ 32 =**

3) **17.4 - 3 x (2.8 - 1.6) =**

4) **(7.25 + 2.75) x 0.39 =**

5) $\frac{2.4+3.2\times 0.5}{3.2-2\times 0.9}=$

6) $\frac{2.5\times 2.5-2.5\times 0.5}{0.7\times 4-2.3}=$

7) $\frac{0.7\times (0.9-0.9\times 0.9)}{0.3\times 0.3}=$

8) $\frac{(4.2-1.4\times 2.5)\times 20}{3.5\div 0.05}=$

Lesson 3
Combined Operations with All Kinds of Numbers

So far, we have worked only with the same kind of numbers at the same time. But is it possible to work with different kinds of numbers: decimals, and fractions, and whole or mixed numbers at the same time? Of course. You already know that we can write any fraction as a decimal.

Unfortunately, not all of them are terminated. You also know how to write a decimal as a fraction. It means that you can solve a problem even it has different kinds of numbers in it. You should be very careful in choosing the way to solve each problem. Sometimes we rewrite all the decimals in the problem as fractions, and sometimes it is more convenient to rewrite all the fractions as decimals and work with the decimals.
Each problem is different and should be solved differently.

For example, $\frac{3}{4}\times 0.4-\frac{2}{3}\times 0.3$.

If we change the decimals into fractions, we can solve this problem easily.

$$\frac{3}{4}\times 0.4-\frac{2}{3}\times 0.3=\frac{3}{4}\times\frac{4}{10}-\frac{2}{3}\times\frac{3}{10}=\frac{3}{10}-\frac{2}{10}=\frac{1}{10}=0.1$$

If we decide to change all the fractions into decimals, we will not know what to do with $0.\overline{6}\left(\frac{2}{3}=0.\overline{6}\right)$, and we will not be able to solve this problem.

To solve this kind of problems easily, you should understand several important things very well. First, that we can write each terminated decimal as a fraction.

For example, $0.73=\frac{73}{100}$, $0.125=\frac{125}{1000}=\frac{1}{8}$, and so on.

Next, that if a fraction has factors of 10, 100, 1000 and so on in the denominator, you can be sure that you can write this fraction as a decimal.

For example, $\frac{1}{4}=\frac{1\times 25}{4\times 25}=\frac{25}{100}=0.25$ or $\frac{3}{125}=\frac{3\times 8}{125\times 8}=\frac{24}{1000}=0.024$

The last important thing is that we can write any whole number as a decimal with zeros after the decimal point or as a fraction with **1** as a denominator.
Now let us solve some problems with all kinds of numbers.

$$\frac{3}{7}\times 0.28+\frac{2}{3}\div\frac{4}{9}=\frac{3}{7}\times\frac{28}{100}+\frac{2}{3}\times\frac{9}{4}=0.12+\frac{3}{2}=0.12+1.5=1.62$$

$$0.25\times 4-3\times\left(1-\frac{3}{4}\right)=0.25\times 4-3\times\frac{1}{4}=1-\frac{3}{4}=\frac{1}{4}=0.25$$

$$\frac{12\times\frac{5}{6}-4\times 0.3}{3.6+\frac{2}{5}\times 2}=\frac{\frac{12\times 5}{1\times 6}-1.2}{3.6+\frac{2\times 2}{5\times 1}}=\frac{10-1.2}{3.6+\frac{4}{5}}=\frac{8.8}{3.6+0.8}=\frac{8.8}{4.4}=2$$

Now try to solve problems by yourself.

PRACTICE

$3\frac{1}{4} - 0.7 \times 0.6 =$

$2.75 + 3 \times \left(\frac{1}{2} - \frac{1}{4}\right) =$

$\frac{3}{5} \times \frac{2}{7} - 0.3 \times \frac{1}{21} =$

$0.24 \div 0.5 - \frac{3}{4} \div \frac{3}{2} =$

$11.2 - 3.6 \times \left(1\frac{2}{3} - \frac{6}{7} \div \frac{9}{14}\right) =$

$7 \times \left(2\frac{6}{7} + 0.3 \div 0.1\right) =$

$\frac{0.3 \div 0.06 - 6\frac{2}{3} \div 3\frac{1}{3}}{2 + \frac{3}{7} \times \frac{7}{3}} =$

$\frac{\frac{3}{8} \times \frac{4}{9} + 2 \times \frac{5}{9}}{0.4 - 0.3 \div 3} =$

Lesson 4
Word Problems

1. Bill is 7 years older than Jim. The sum of their ages is 43. How old are they?

2. The Big Bad Wolf is chasing a little pig who is running away at 5 m/s. The wolf can run at 7 m/s. The initial distance between them is 50 m. If the little pig's house is 250 m away, will the pig be able to escape? Justify your answer.

3. A frog and a toad swim toward each other from opposite sides of a pond. The frog can swim at 3 m/s, and the toad at 4 m/s. If they meet in 6 seconds, how wide is the pond?

4. Ben and Eric are two candidates for the class president. Ben received $\frac{3}{5}$ of the votes, and Eric received 12 votes. How many students are in the class?

5. Three girls went to a grocery store. They had \$9 all together. Two of them had equal amounts of money, and the third girl had twice as much as each of her friends. How much money did each girl have?

6. Daniel was doing his math homework. At first, he solved 7 easy problems and spent 3 minutes on each of them. Then he solved 5 more difficult problems, and spent 7 minutes on each of them. At the end, he could not solve 2 very tough problems in spite of the fact that he spent 25 minutes on each. How long did it take him to do his math homework?

7. Two friends decided to meet each other and left their houses at the same time. Josh ran with a speed of 5 m/s, but Jared walked slowly with a speed of 2 m/s. When will they meet each other if the distance between their houses is 637 meters?

8. Mary lost 21 mittens during the winter. Her unlucky mother found only one third of them. How many mittens were lost forever?

9. Gwen's grandmother is twice older than Gwen's mother, but Gwen is three times as young than her mother. What is the sum of their ages if the grandmother is 66 years old?

10. The Wilson family went on vacation. The first day, they spent half of their money; the next day they spent a half of what was left, and on the third day they spent the last $200. How much money did they have at the beginning of their vacations?

Chapter 10
Integers

Lesson 1
Integers, Number Line

The smallest number you know is 0, which represents nothing. However, as strange as it may sound, there are numbers that are less than nothing. How can this be? You might have noticed that a thermometer's scale doesn't end in zero – it can be even colder than 0 degrees. Also, it is possible to have less than no money – if you have nothing in the bank and you owe a sum of money, you have less than nothing. To express such situations mathematically, people use **negative numbers** – numbers that are less than 0.

Look at the number line. All numbers to the right of 0 are called positive numbers.
All numbers to the left of 0 are called negative numbers.
We will work only with whole numbers; all whole numbers in both directions away from zero are called **integers.**

negative integers zero positive integers

-5 -4 -3 -2 -1 0 1 2 3 4 5

Zero is neither positive nor negative; it is neutral.

The **absolute value of a number** is the number of units the integer is away from 0 on the number line. **Whether a number is positive or negative, its absolute value is always a positive number.**

The absolute value of 5 is written | 5 |.

5 is 5 units away from 0, i.e. | 5 | = 5

-5 is also 5 units away from 0, i.e. |- 5 |= 5

Numbers with the same absolute value but different signs are called **opposites.**
You can see that 5 and -5 are opposites.

We can say that 7 is the opposite of -7, and -12 is the opposite of 12.

The location of each point can be defined easily if it is placed on the number line.

A B C -1 0 1 D E F

You can see that points A, B, and C are negative integers, but D, E, and F are positive.
A (-6), B (-4), C (-2), D (3), E (5), F (6).

Points A and F are the same distance away from 0; that means that their absolute values are the same.

| -6 | = | 6 | = 6

It means that they are opposites.

Now practice.

PRACTICE

Give the absolute value.

1. | -13 | 2. | 23 | 3. | -122 | 4. | -41 | 5. | 68 |

Give the opposites of these numbers.

6. -11 7. 99 8. -36 9. -52 10. 106

Give the integer for each point on the number line.

K L M -1 0 1 N P Q

11. K 12. L 13. M 14. N 15. P 16. Q

Lesson 2
Comparing Integers

We will use the same rule for comparing integers as we used for comparing whole numbers. The farther an integer is to the right on the number line, the greater it is. It means that -2 is greater than -5, or we can write it like this: -2 > -5

smaller____.____.____.____.____.____.____|____.____.____.____.____.____._greater
-6 -5 -4 -3 -2 -1 0 1 2 3 4 5 6

You can think about it like this: the more you owe somebody, the poorer you are. It is better to owe 2 dollars than 30 dollars. Or, it is warmer if the temperature is -8 degrees that if it is -32.
You can see that any positive number is greater than any negative number.
When we compare negative numbers, we can say that a number with the smaller absolute value is greater than a number with the greater absolute value.

- 15 < -7 -4 < 2 3 > -5 0 > -6 | - 8 | > 5 -8 < 5

To order a set of integers, we should compare each integer to the others in turn.
For example, let us order the numbers: -11, 5, 0, -3, 12, -24, | -36 |
-24 is the farthest to the left on the number line, -11 is next, then -3, then 0, 12 is next. The absolute value of any number is positive, which is why | -36 | is the greatest number; it is farthest to the right.
So, -24 < -11 < -3 < 0 < 5 < 12 < | -36 |

PRACTICE
Use the symbols < or > to order the numbers.
2 and -2 -12 and | -13 | -7 and -9 8 and 14

Use the symbols < or > to order from least to greatest

-12, 7, -20 | -6 |, 5, -4, 0, -3 -18, 3, 0, -25, -2

Use the symbols < or > to order from greatest to least

8, -18, | -28 |, 14 1, -16, -8, | 9 |, | -12 |

Lesson 3
Adding Integers

To understand the idea of adding integers, we will use the number line. Remember that positive numbers go to the right of zero, but negative numbers go to the left of zero on the number line.

-8 -7 -6 -5 -4 -3 -2 -1 0 1 2 3 4 5 6 7 8

To add numbers, imagine a little inchworm crawling along the number line. If it starts at 0, makes 3 steps to the right, and then 5 steps more, it will find itself at 8. We can say that we added 3 and 5 and got 8. 3 + 5 = 8.

Now suppose we want to add two negative integers, (- 2) + (- 4). Now, our inchworm will be crawling in the negative direction (to the left). If it makes two steps to the left and then makes 4 more steps to he left, it will end up at (-6). It is understandable; if you keep going in the same direction, you will end up farther to the left. You can also think this way about adding negative numbers. If you already had a debt and then borrowed more money, you will, of course, owe even more.

Now I want to give you a rule for adding integers with the same signs (or we can say with like signs).

1. Add their absolute values.
2. Give the sum the sign of the integer addends.

Let us solve several problems, using the number line or this rule.

(-5) + (-3) = (- 8)

1) Find (- 5) on the number line and make 3 steps to the left. You are at (- 8).

2) Using the rule, you will get the same answer.

|- 5 | + | -3| = 8. Since both addends are negative, their sum is negative too .

12 + 4 = 16 (- 7) + (- 21) = (- 28) 6 + 24 = 30

But how can we add numbers with different or unlike signs? Again, let us use the number line. Suppose we want to add 5 and (-7). We will do the same things: at first our inchworm will move 5 steps to the right of 0 and then make 7 steps to the left, because the second addend is negative. It ends up at (-2). It means that 5 + (- 7) = (- 2).

Another example: (- 6) + 9. Move 6 steps to the left and then 9 steps to the right (the second addend is positive) and you will end up at 3. It means that (- 6) + 9 = 3

Unfortunately, we can't always use the number line to calculate. It is very convenient for small numbers, but what if we have to calculate (-137) + 89? We can draw a very very long number line, or we can save paper and ink and try to find another way to figure this out. Use the number line to calculate the answer to 5 + (-2), and compare it to the answer to 5 – 2. The two answers should be the same. If you think about it, it makes sense – in both cases, the inchworm will move 5 steps to the right and 2 steps to the left. So, we see that subtracting an integer is the same thing as adding its opposite. This can help us to formulate the rules for adding integers with unlike signs.
Here is a rule for adding integers with unlike signs.

1. Find the difference of their absolute values.
2. Give the difference the sign of the integer addend with the greater absolute value.

Look at these examples.

4 + (- 7) = (- 3)
1) Find 4 on the number line , then make 7 steps to the left. You are at (- 3).
2) Now use the rule: | - 7 | - | 4 | = 3. Since | - 7 | is larger than | 4 |, the sum is negative .

(- 8) + 12 = 4 { | 12 |- | - 8 | = 4. Since | 12 | is larger than | - 8 |, the sum is positive }

I think that you understand that the sum of any integer and its opposite is zero!

-5 + 5 = 0 12 + (- 12) = 0
If the difference is zero, we will omit the sign. Zero can't be negative or positive.

PRACTICE

9 + (- 12) =	(- 3) + (- 14) =	(- 15) + 8 =	22 + 13 =
(-22) + 8 =	16 + (- 7) =	(- 42) + (- 18) =	32 + (-32)=
17 + 35 =	(- 11) + (- 24) =	14 + (- 29) =	(- 62) + 42 =

Lesson 4
Subtracting Integers

To understand how to subtract integers, you will use the number line again. Actually, you will do the same things as you did when you subtracted whole numbers. But now, remember that negative integers go in the opposite direction.

-8 -7 -6 -5 -4 -3 -2 -1 0 1 2 3 4 5 6 7 8

Suppose we want to subtract 2 from 5. We should find 5 on the number line, then make 2 steps to the left. We are at point 3, i.e. 5 - 2 = 3. You knew it, didn't you?

But with integers, you can subtract a bigger number from a smaller one – for example, 4 - 7. Again start by finding 4 on the number line, then make 7 steps to the left, and you are at point -3. It means that 4 - 7 = (-3).
You probably noticed that when we subtract, we go in the direction that is opposite to that of addition. Therefore, when we subtract a negative number, we should go to the right.
For example, (-3) - (-8). Find -3 on the number line and then make 8 steps to the right, and you are at point 5. It means that (-3) - (-8) = 5.
You might also have noticed that when you subtract a positive number, you do the same as when you add a negative number: you make steps to the left. When you subtract a negative number, you do the same as when you add a positive number: you make steps to the right. This observation gives us a very convenient rule for subtracting integers:

To subtract an integer, add its opposite.

At first, you can see that it works with positive numbers.

8 - 3 = 8 + (-3) = 5   12 - 6 = 12 + (-6) = 6

Now, let us work with integers.

-7 - (-3) = -7 + 3 = (-4) (instead of subtracting -3, we added the opposite of –3, which is 3). You can check it on the number line. First, find –7; then make 3 steps to the right, because you are subtracting a negative number, and you will reach the point (-4).

Look at these examples:

- 6 - 5 = - 6 + (- 5) = (- 11)   5 - 13 = 5 + (- 13) = (- 8)

- 12 - (- 7) = - 12 + 7 = (- 5)   18 - (- 22) = 18 + 22 = 40

PRACTICE

Subtract:

- 4 - 7 =	8 - 13 =	-7 - (- 8) =	- 20 - (- 9) =
15 - 6 =	-24 - 25 =	18 - (- 16) =	-14 - 24 =
22 - 22 =	13 - (- 13) =	48 - 26 =	2 - 17 =

Lesson 5
Multiplying Integers

Do you remember the meaning of multiplication? I think you do. When we have to add the same number several times, we use multiplication. We will do the same with integers. Suppose we

have to take (-3) seven times. We will write it like this: (-3) x 7, but it really means this:

$(-3)+(-3)+(-3)+(-3)+(-3)+(-3)+(-3)=(-21)$

You can see that if you multiply a negative number by a positive number, you will get a negative number. The absolute value of the product will be the product of the absolute values of both factors.
We have already talked about the commutative property of multiplication; therefore, we can say that the result of multiplication of a positive number by a negative number is negative.

But what happen if we multiply a negative number by a negative number?
Let us think. If we multiply a positive number by a positive number we got a positive result. Then we change the sign of one of the factors and got a negative result, or the opposite of the first one. If we change the sign of the second factor, we will get the opposite of the second result, or, in other words, the result of the multiplication of a negative number by a negative number is positive.

This is the rule of multiplying integers:

1) The product of two integers with like signs is positive.
2) The product of two integers with unlike signs is negative.
3) The absolute value of the product equals the product of the absolute values of both factors.

Look at these examples.

8 x (- 4) = (- 32) { positive times negative is negative }

(- 5) x (- 12) = 60 { negative times negative is positive }

(- 7) x 7 = (-49) { negative times positive is negative }

(- 2) x (- 3) x (- 4) = (- 24) { negative times negative is positive, positive times negative is negative }

PRACTICE

EXERCISES

9 x (- 5) = (- 6) x (- 12) = 18 x 7 = (- 20) x 5 =

(- 3) x 4 x (- 8) = (- 1) x (- 2) x (- 3) x (- 4) = 7 x (- 6) x 5 =

Lesson 6
Dividing Integers

To find the rules of dividing integers, let us recall the relationship between multiplication and division. Let us look at this example: 6 x 7 = 42. But we know that 42 ÷ 7 = 6 and 42 ÷ 6 = 7. It means that if we make the product into a dividend, and one of the factors into a divisor, the second factor will be the quotient. We know that if one of the factors is negative, the product is negative. So, we can conclude that if we divide a negative number by a positive number, the quotient will be negative; but if we divide a negative number by a negative number, the quotient is positive.

Look at this example: 3 x (-8) = (-24) , i.e.(-24) ÷ 3 = (-8) and (-24) ÷ (-8) = 3

We know that if both factors are negative, the product is positive. So, we can say that if we divide a positive number by a negative one, we will get a negative answer.

Look at this example: (-5) x (-6) = 30, i.e. 30 ÷ (-6) = (-5)

I should not tell you, because you already know it perfectly well, that if we divide a positive number by a positive one, we will get a positive answer.

These are the rules for dividing integers:

1) The quotient of two integers with like signs is positive
2) The quotient of two integers with unlike signs is negative
3) The absolute value of the quotient is the result of division of the absolute values of a dividend and a divisor

Now look at several examples

25 ÷ (-5) = (-5) { positive divided by negative is negative }

(-24) ÷ (-6) = 4 { negative divided by negative is positive }

(-35) ÷ 7 = (-5) { negative divided by positive is negative }

40 ÷ 4 = 10 { positive divided by positive is positive }

PRACTICE

56 ÷ (- 7) = (- 42) ÷ (- 6) = 28 ÷ 7 = (- 72) ÷ 12 =

100 ÷ (- 10) = (- 64) ÷ 4 = 0 ÷ (-13) = 96 ÷ (- 16) =

Lesson 7
Combined Operations With Integers

Do you remember the order of operations? If not, review Chapter 2, Lesson 1.
When you studied that material, you didn't know integers. Now you can evaluate these expressions even if they have integers, too. And here is some good news: the order of operations is the same. It means that, again, we will do multiplication and division first, and then addition and subtraction.
If we have to change the order, we will use parentheses, ( ), or brackets, [ ].
If we have operations of the same priority, we will do them in order from left to right.

For example: (-7) + 3 x (-2).
At first, we will do multiplication: 3 x (-2) = (-6)
The next step is addition: (-7) + (-6) = (-13)

Write it like this: (- 7) + 3 x (- 2) = (- 7) + (- 6) = (- 13)

Another problem: 4 x (- 5) - (- 6) x 7 = (- 20) - (- 42) = (- 20) + 42 = 22
At first we multiplied 4 x (- 5), then (- 6) x 7; and then we subtracted the second product from the first one.

Now look at this problem with parentheses and brackets. With integers, we mostly use parentheses to write negative numbers and brackets to separate out operations; we will do the operation inside the brackets first.
12 + (- 3) x [7 - (-2)]
1) 7 - (-2) = 7 + 2 = 9
2) (- 3) x 9 = (-27)
3) 12 + (- 27) = (- 15)

Or you can write it this way: 12 + (- 3) x [7 - (- 2)] = 12 + (- 3) x 9 = 12 + (- 27) = (- 15)

Another example: (- 12) ÷ 6 - 8 x (-2)
At first do the division, then multiplication, then subtraction.

(- 12) ÷ 6 - 8 x (- 2) = (- 2) - (- 4) = (- 2) + 4 = 2

We will do more complicated calculations at the end of this book; now practice.

PRACTICE

1) 17 - 5 x (- 4) =

2) 9 x (- 2) + (- 6) ÷ (- 3) =

3) 7 x (8 - 12) - 15 =

4) (- 11 - 9) x 3 + (- 4) x (2 - 6) =

5) 24 ÷ (- 6) - 18 ÷ 3 + (- 36) ÷ 6 =

6) 13 + (- 2) x [20 - (- 15)] =

7) (- 18) - (- 16) + (- 14) ÷ 2 =

8) 100 ÷ [(- 20) ÷ (3 - 5)] =

9) [80 ÷ (-40) –4] - 16 =

10) (- 50) ÷ [50 - (50 - 75) + (- 25)] =

Lesson 8
Word Problems

1. A student earned 5 points extra credit for solving a difficult word problem. Next day, his teacher deducted 10 points because he didn't do his homework. What total number of points did the student get?

2. A family flew from Texas to Minnesota. The temperature in Texas was 20 C; in Minnesota it was -12 C. What difference in temperature will the family experience?

3. Tom scored - 8 in the first round of a game, and +6 in the second round. Find his score after two rounds.

4. Jacob had $ 825 in his checking account. He wrote a check for $909. What is his new balance?

5. An airplane is flying 3 km above the ground. The temperature on the ground is 0C. What is the temperature outside the airplane, if the air temperature decreases 7C for each kilometer of increase in altitude?

6. At 5 a.m. the temperature was -7 C, but at noon it was 12 C. How many degrees did the temperature rise?

7. An elevator starts at the 24th floor, goes down 18 floors and then up 6 floors. At what floor is it then?

8. A boy jumped from a diving board into a swimming pool. What distance has he covered if the diving board is 5m above the water and the swimming pool is 3m deep? Assume he goes all the way to the bottom of the pool.

9. After a flood, the water level of a river decreased 5 inches a day. If during the flood, the river was 30 inches higher than normal, how many days will it take for the water level to decrease to 15 inches below normal?

10. Between the early morning and noon the temperature increased 42 F and reached 29 F at noon. What was the temperature in the morning?

Chapter 1

Lesson 1 Homework 1

MULTIPLY

$\begin{array}{r} 20 \\ \times\ 30 \\ \hline \end{array}$	$\begin{array}{r} 40 \\ \times\ 50 \\ \hline \end{array}$	$\begin{array}{r} 70 \\ \times\ 80 \\ \hline \end{array}$	$\begin{array}{r} 900 \\ \times\ 6 \\ \hline \end{array}$
$\begin{array}{r} 8 \\ \times\ 300 \\ \hline \end{array}$	$\begin{array}{r} 50 \\ \times\ 40 \\ \hline \end{array}$	$\begin{array}{r} 70 \\ \times\ 80 \\ \hline \end{array}$	$\begin{array}{r} 90 \\ \times\ 9 \\ \hline \end{array}$
$\begin{array}{r} 5 \\ \times\ 40 \\ \hline \end{array}$	$\begin{array}{r} 80 \\ \times\ 30 \\ \hline \end{array}$	$\begin{array}{r} 3 \\ \times\ 40 \\ \hline \end{array}$	$\begin{array}{r} 80 \\ \times\ 60 \\ \hline \end{array}$
$\begin{array}{r} 600 \\ \times\ 400 \\ \hline \end{array}$	$\begin{array}{r} 80 \\ \times\ 200 \\ \hline \end{array}$	$\begin{array}{r} 30 \\ \times\ 90 \\ \hline \end{array}$	$\begin{array}{r} 7{,}000 \\ \times\ 7 \\ \hline \end{array}$
$\begin{array}{r} 50 \\ \times\ 4 \\ \hline \end{array}$	$\begin{array}{r} 3 \\ \times\ 800 \\ \hline \end{array}$	$\begin{array}{r} 60 \\ \times\ 300 \\ \hline \end{array}$	$\begin{array}{r} 600 \\ \times\ 40 \\ \hline \end{array}$

Lesson 2 Homework 1

MULTIPLY

24 × 2	32 × 3	14 × 4	25 × 2	41 × 5
22 × 4	43 × 5	51 × 6	84 × 3	18 × 5
74 × 5	92 × 4	68 × 6	37 × 7	56 × 8
83 × 7	60 × 6	59 × 5	96 × 9	29 × 4
61 × 8	87 × 7	44 × 9	58 × 6	19 × 3

Lesson 2 Homework 2

MULTIPLY

28 × 5	63 × 4	45 × 6	72 × 8	81 × 9
57 × 8	34 × 6	60 × 7	93 × 5	48 × 3
14 × 7	52 × 8	34 × 4	21 × 9	77 × 2
62 × 6	84 × 5	90 × 8	56 × 3	73 × 4
28 × 7	59 × 2	68 × 0	71 × 8	96 × 5

Lesson 3 Homework 1

MULTIPLY

356 × 5	297 × 8	612 × 7	408 × 9	574 × 3
493 × 3	280 × 4	175 × 6	788 × 9	306 × 2
916 × 7	273 × 5	492 × 2	708 × 4	526 × 8
870 × 9	169 × 4	747 × 6	952 × 3	638 × 7
126 × 5	864 × 8	357 × 3	905 × 6	774 × 2

Lesson 3 Homework 2

MULTIPLY

236 x 7	842 x 9	567 x 6	971 x 4	807 x 3
593 x 2	285 x 8	609 x 5	173 x 9	427 x 6
477 x 3	582 x 7	729 x 8	641 x 4	807 x 5
397 x 5	411 x 6	699 x 7	708 x 9	543 x 3

Lesson 3 Homework 3

MULTIPLY

134 x 7	2,356 x 3	34,507 x 8
5,608 x 6	29,720 x 9	9,335 x 4
62,677 x 3	380,641 x 7	8,075 x 2
4,603,542 x 5	56,873,021 x 8	890,671,532 x 6
70,865,732 x 4	59,704,512 x 7	9,480,443 x 9

Lesson 3 Homework 4

MULTIPLY

368 x 6	937 x 8	1,504 x 9
2,570 x 3	32,708 x 7	65,411 x 5
27,802 x 9	378,123 x 4	609,502 x 8
3,708,223 x 6	4,120,389 x 7	1,234,567 x 8
203,405,607 x 9	311,287,460 x 5	293,445,891 x 2

Lesson 4 Homework 1

DIVIDE

$3\overline{)54}$	$4\overline{)76}$	$2\overline{)94}$
$5\overline{)65}$	$6\overline{)84}$	$7\overline{)91}$
$8\overline{)96}$	$9\overline{)99}$	$3\overline{)81}$
$4\overline{)92}$	$7\overline{)84}$	$6\overline{)96}$

Lesson 4 Homework 2

DIVIDE

$2\overline{)68}$	$5\overline{)75}$	$3\overline{)78}$
$4\overline{)56}$	$6\overline{)84}$	$8\overline{)88}$
$7\overline{)98}$	$9\overline{)90}$	$5\overline{)90}$
$6\overline{)75}$	$7\overline{)82}$	$4\overline{)61}$
$2\overline{)39}$	$3\overline{)58}$	$8\overline{)93}$

Lesson 5 Homework 1

DIVIDE

$3\overline{)135}$ $4\overline{)524}$ $5\overline{)275}$

$2\overline{)776}$ $6\overline{)834}$ $9\overline{)261}$

$7\overline{)735}$ $8\overline{)336}$ $5\overline{)680}$

$6\overline{)996}$ $9\overline{)477}$ $7\overline{)815}$

$2\overline{)537}$ $3\overline{)431}$ $4\overline{)625}$

Lesson 5 Homework 2

DIVIDE

$4\overline{)2{,}578}$ $5\overline{)46{,}395}$

$6\overline{)783{,}455}$ $7\overline{)359{,}258}$

$8\overline{)567{,}205}$ $9\overline{)13{,}022}$

$3\overline{)45{,}103}$ $2\overline{)33{,}414{,}866}$

$6\overline{)273{,}104}$ $4\overline{)28{,}337{,}242}$

Lesson 5 Homework 3

DIVIDE

$2\overline{)356{,}103}$

$3\overline{)452{,}072}$

$4\overline{)783{,}235{,}064}$

$5\overline{)67{,}545{,}325}$

$6\overline{)1{,}234{,}567{,}890}$

$7\overline{)89{,}663{,}208}$

$8\overline{)24{,}345{,}621{,}046}$

$9\overline{)46{,}678{,}075{,}077}$

$7\overline{)7{,}678{,}550{,}099}$

$6\overline{)3{,}658{,}775{,}067}$

Lesson 6 Homework 1

MULTIPLY

56 x 20	74 x 30	38 x 40
92 x 24	63 x 32	54 x 63
81 x 55	19 x 47	27 x 72
302 x 26	205 x 37	601 x 76
283 x 31	711 x 56	829 x 42

Lesson 6 Homework 2

MULTIPLY

78 x 50	65 x 60	93 x 70
34 x 28	52 x 53	48 x 67
354 x 50	672 x 70	926 x 80
506 x 35	908 x 27	344 x 48
692 x 56	834 x 72	750 x 69

Lesson 6 Homework 3

MULTIPLY

29 × 56	68 × 49	93 × 72
358 × 45	189 × 67	563 × 39
2,409 × 63	4,380 × 58	5,227 × 94
42,781 × 59	76,902 × 67	78,233 × 84
710,663 × 82	324,129 × 64	458,014 × 79

Lesson 7 Homework 1

MULTIPLY

38 x 400	456 x 300	1,272 x 900
201 x 135	713 x 206	494 x 322
1,320 x 205	2,451 x 312	5,821 x 480
8,369 x 422	6,708 x 516	3,473 x 193
12,234 x 516	405,882 x 734	672,089 x 295

Lesson 7 Homework 2

MULTIPLY

202 × 700	421 × 900	390 × 250
165 × 214	592 × 608	853 × 175
1,508 × 412	3,821 × 637	7,922 × 508
70,245 × 388	62,903 × 721	45,824 × 906
345,806 × 217	703,924 × 435	558,067 × 924

Lesson 8 Homework 1

DIVIDE

$11\overline{)264}$ $13\overline{)416}$ $21\overline{)672}$

$17\overline{)901}$ $23\overline{)989}$ $32\overline{)544}$

$14\overline{)504}$ $41\overline{)984}$ $38\overline{)1,330}$

$52\overline{)3,016}$ $49\overline{)1,813}$ $19\overline{)532}$

$35\overline{)1,435}$ $61\overline{)2,928}$ $29\overline{)1,943}$

Lesson 8 Homework 2

DIVIDE

$16\overline{)560}$	$22\overline{)1{,}166}$	$34\overline{)2{,}074}$
$43\overline{)1{,}204}$	$63\overline{)2{,}016}$	$80\overline{)4{,}560}$
$37\overline{)929}$	$54\overline{)1{,}800}$	$30\overline{)818}$
$78\overline{)4{,}140}$	$62\overline{)2{,}760}$	$40\overline{)3{,}688}$
$92\overline{)3{,}500}$	$88\overline{)1{,}344}$	$71\overline{)2{,}100}$

Lesson 8 Homework 3

DIVIDE

$12\overline{)1{,}608}$	$21\overline{)2{,}373}$	$15\overline{)1{,}320}$
$30\overline{)6{,}420}$	$42\overline{)8{,}946}$	$51\overline{)15{,}402}$
$17\overline{)2{,}100}$	$28\overline{)5{,}888}$	$23\overline{)5{,}437}$
$99\overline{)19{,}899}$	$70\overline{)15{,}762}$	$34\overline{)7{,}040}$
$56\overline{)8{,}240}$	$82\overline{)19{,}790}$	$60\overline{)19{,}980}$

Lesson 9 Homework 1

Find the average of these numbers:

1. 2, 5, 11
2. 4, 19, 15, 2
3. 38, 48
4. 3, 4, 7, 6, 5
5. 12, 15, 22, 11, 0
6. 30, 40, 50, 60
7. 121, 175, 124
8. 89, 148, 203, 120
9. 7, 18, 29, 14, 32, 20
10. 326, 305, 314

Word problems

11. Ben got 88% on his first math test, 94% on the second one, and only 82% on the third one. What was his average in math after these three tests?

12. Susan read a very interesting book in 4 days. She read 87 pages the first day, 93 the second, 63 the third, and 77 the last day. What was the average number of pages that she read each day?

13. Rick's salary is $18,000 per year. His manager gets twice as much. What is their average salary?

14. The Davidsons went on a trip for a long weekend. On the first day they drove 360 miles, the second day only 150 miles, the third day 280, and the last day 310 miles. What was the average distance per day?

Lesson 10 Homework 1

Word problems

1. The Jackson family watches TV 42 hours per week. How many hours do they watch TV during a year? (hint: one year has 52 weeks).

2. On his birthday Max bought 432 candies for his classmates. How many candies did each kid eat if there are 18 kids in the class, and Max divided the candies equally?

3. Every day Sam got three A's and two B's at school. How many A's and B's did he get for the whole year, if children spend 175 days in school?

4. Mr. Brown, the owner of a local sock store in a small town, noticed that each person in town buys four pairs of socks each year. He knows that at the beginning of the year he ordered 880,000 pairs of socks, at the end of the year he has only 8,000 pairs. How many people live in the town?

5. Three brothers worked in a grocery store to save money for college. They earned $3,900 altogether. After the oldest brother took $200 and the middle brother took $100 from this amount, they divided the money equally. How much did each get?

6. Two best friends, Tim and Tom, had a big collection of baseball cards: 320 cards altogether. Tim had to move to a different state, and Tom divided their collection between the two of them. He gave 40 cards more to Tim than to himself. How many cards does each boy have?

7. An old math teacher at the end of her career realized that she checked 23,460 tests. How much time did she spend checking tests if she spent about 12 minutes for each?

8. One day a policeman stopped 17 drivers for speeding. Each driver paid a 117 dollar fine. How much did they pay altogether?

9. How many minutes are in one day? in one month ? in one year? (hint: 1 day = 24 hours, 1 month = 31 days, 1 year = 365 days)

10. A class went to see a play in a theater. The teacher paid $672 for the tickets. How much does each ticket cost if there are 27 students in the class and only one teacher was with them?

Chapter 1 Test

$$\begin{array}{r} 28 \\ \times\ \underline{7} \end{array} \qquad \begin{array}{r} 317 \\ \times\ \underline{8} \end{array} \qquad \begin{array}{r} 924 \\ \times\ \underline{3} \end{array} \qquad \begin{array}{r} 2{,}568 \\ \times\ \underline{2} \end{array}$$

$$\begin{array}{r} 76 \\ \times\ \underline{43} \end{array} \qquad \begin{array}{r} 619 \\ \times\ \underline{52} \end{array} \qquad \begin{array}{r} 246 \\ \times\ \underline{79} \end{array} \qquad \begin{array}{r} 3{,}261 \\ \times\ \underline{408} \end{array}$$

$$2\,\overline{)\,74} \qquad 3\,\overline{)\,57} \qquad 4\,\overline{)\,612} \qquad 5\,\overline{)\,385}$$

$$11\,\overline{)\,374} \qquad 24\,\overline{)\,816} \qquad 52\,\overline{)\,1{,}352} \qquad 43\,\overline{)\,2{,}408}$$

WORD PROBLEMS

1. The Davidsons have 3 apple trees in their back yard. In the fall one tree had 112 apples, the second one had 287, while the third one had only 12. What was the average amount of apples on each tree?

2. Nathan missed only 2 days during a school year in his school. Every day he had 8 lessons. How many lessons did Nathan have during the school year? (hint: a school year is 175 days long).

3. A person uses approximately 180 per hour calories playing table tennis and 250 calories dancing. On Sunday Liora had a dancing class for 3 hours, and then she played tennis with her father for 2 hours. How many calories did she spend on these activities?

Chapter 2

Lesson 1 Homework 1

Evaluate the following expressions.

1. $12 + 3 \times 4 =$
2. $(12 + 3) \times 4 =$
3. $8 \times 3 + 5 \times 6 =$
4. $8 \times (3 + 5) \times 6 =$
5. $16 - 16 \div 4 =$
6. $(24 - 8) \div 4 =$
7. $13 + 3 \times (18 - 7) =$
8. $64 \div (17 - 9) + 12 =$
9. $6 + 2 \times [\ 13 - 2 \times (7 - 4)\ ] =$
10. $[\ 18 \div 3 - 5 \times (4 - 3) + 7\ ] \times 2 =$
11. $\dfrac{2 \times 3 + 4 \div 2}{7 - 3 \times (2 - 1)} =$
12. $\dfrac{24 - 3}{7} - \dfrac{3 + 3 \times (4 - 2)}{5 - 2} =$
13. $[\ 1000 \div (500 \div 20)\ ] - 20 =$
14. $1000 \div [\ 200 \div (12 - 7) - 30\ ] =$
15. $\dfrac{24 - 6 \times 3 + 2 \times 2}{10 \div 2} =$

Lesson 1 Homework 2

Evaluate the following expressions.

1. $17 + 3 \times 5 =$
2. $(26 - 18) \div 4 =$
3. $20 - 20 \div 5 =$
4. $7 \times 4 - 3 \times (7 - 5) =$
5. $45 \div 5 - 4 \times 2 =$
6. $16 \div (8 \div 2) \times 7 =$
7. $25 - (15 + 10) \div 5 =$
8. $30 \div (3 \times 2 + 4) =$
9. $2 \times [18 - 3 \times (4 - 2 \times 2)] =$
10. $15 - [6 \times 4 - 5 \times (6 - 4)] \div 7 =$
11. $\dfrac{64 \div 8 - 3 \times 2}{12 \div (6 \times 2)} =$
12. $\dfrac{38 - 4 \times (3 \times 3)}{4 - 4 \div 2} + \dfrac{35}{7} =$
13. $120 \div [250 \div (75 - 25) + 35 \div 5] =$
14. $3 \times [20 \times (3 + 2) - 15 \div (2 + 3)] =$
15. $\dfrac{5 + 2 \times (7 - 2)}{3 \times 3 - 2 \times 2} - \dfrac{12 - 6 \div 3}{4 + 3 \div 3} =$

Lesson 2 Homework 1

Round to the nearest ten.

1. 28 2. 34 3. 265

Round to the nearest hundred.

4. 172 5. 2,709 6. 3,448

Round to the nearest thousand.

7. 7,500 8. 381,234 9. 38,888

Round to the nearest ten thousand.

10. 22,222 11. 4,555,203 12. 496,321

Round to the nearest hundred thousand.

13. 78,772,429 14. 2,345,678

Round to the nearest million.

15. 24,528,909 16. 661,099,367

Round to the nearest ten million.

17. 336,221,903 18. 2,884,728,120

Lesson 2 Homework 2

Round to the nearest ten.

1. 156
2. 89
3. 2,313

Round to the nearest hundred.

4. 4,352
5. 12,345
6. 961

Round to the nearest ten thousand.

7. 72,889
8. 336,667
9. 2,304,449

Round to the nearest hundred million.

10. 67,562,339,832
11. 4,448,362,392

Round to the nearest billion.

12. 23,567,089,324
13. 749,721,109,225

Round each number to the greatest place possible.

14. 45,284
15. 345,678,901,234

Round to the place of the underlined digit.

16. 23$\underline{4}$,432,568
17. 90,2$\underline{0}$9,378,773
18. 3,$\underline{4}$78,092
19. $\underline{1}$92,892,456,337

Lesson 3 Homework 1

Estimate the sums and differences.

Round to the nearest ten:

1. 38 + 44 2. 342 - 35 3. 129 + 33

Round to the nearest thousand:

4. 48,523 - 5,297 5. 358,923 + 61,487

Round to the nearest million:

6. 34,567,888 - 3,218,397

7. 8,533,210 + 12,098,873

Use mental math to find the products.

8. 30 x 300 9. 200 x 80 10. 500 x 600

11. 10 x 20 x 30 12. 4,000 x 8,000

Estimate by rounding. Round so you can use mental math.

13. 38 x 9 14. 249 ÷ 48 15. 189 x 11

16. 3,021 ÷ 61 17. 99 x 98 x 102

Lesson 3 Homework 2

Estimate the sums and differences

Round to the nearest hundred:

1. 345 + 554 2. 888 - 119 3. 1,629 - 572

Round to the nearest ten thousand:

4. 56,834 + 23,359 + 18,746

5. 127,732 + 48,214 - 22,190

Round to the nearest billion:

6. 3,482,968,556 + 2,888,374,904

7. 18,723,742,033 - 7,262,870,931

Use mental math to find the products.

8. 700 x 6 9. 2,000 x 40 10. 800 x 900

11. 10 x 200 x 300 12. 400 x 5,000 x 30

Estimate by rounding. Round so you can use mental math.

13. 99 x 11 14. 6,006 x 492

15. 921 ÷ 89 16. 59 x 69 x 71

17. 239 ÷ 8 18. 707 x 696 x 9

Lesson 4 Homework 1

Use front-end digits to estimate.

1. 244 + 549
2. 3,622 + 4,278
3. 3,328 + 4,332 + 5,341
4. 767 - 258
5. 16,532 - 9,499
6. 68 + 43 + 79
7. 411 + 377 - 255

Estimate to find the best missing numbers for each □

8. Missing numbers: 24, 38, 67

a) 32 + □ is close to 100

b) 27 + □ is close to 50

c) 34 + □ is close to 70

9. Missing numbers: 8, 19, 42, 55, 73, 86

a) 49 + □ + □ is close to 100

b) 20 + □ + □ is close to 110

c) 11 + □ + □ is close to 150

Lesson 5 Homework 1

Find the answer in two steps using compensation.

Example: 39 + 67 = (40 + 67) - 1 = 107 - 1 = 106

1. 56 + 9 =
2. 64 - 8 =
3. 48 + 26 =
4. 76 - 19 =
5. 99 + 16 =
6. 67 - 48 =
7. 253 + 398 =
8. 638 - 199 =
9. 677 + 97 =
10. 248 - 97 =
11. 485 + 199 =
12. 555 - 298 =
13. 1,743 + 297 =
14. 2,064 - 96 =
15. 6,246 + 1,996 =
16. 3,741 - 2,995 =
17. 39 x 9 =
18. 45 x 11 =
19. 78 x 99 =
20. 68 x 999 =

Lesson 6 Homework 1

Substitute compatible numbers to estimate an answer.

1. 28 + 73
2. 73 - 26
3. 177 + 126
4. 387 - 193
5. 19 x 6
6. 78 x 11
7. 28 + 54 + 48
8. 147 - 22 - 32

Use distributive property and mental math to find the products

9. 3 x 37 =
10. 4 x 56 =
11. 7 x 27 =
12. 6 x 34 =
13. 66 x 5 =
14. 124 x 2 =
15. 18 x 8 =
16. 81 x 9 =
17. 321 x 3 =
18. 104 x 5 =
19. 82 x 4 =
20. 63 x 7 =

Lesson 7 Homework 1

Word problems. Solve by estimating.

1. The Jenkins family bought a new computer. They paid $999 for the computer, $497 for the monitor, $288 for the printer, and $98 for the scanner. How much did they pay?

2. Seven friends went to see a baseball game. How much did they pay for the tickets if each ticket costs $28?

3. A photographer came to school to take pictures of the kids. He usually spends 19 minutes for each picture. How much time did he spend, if there are 157 children in the school?

4. Michael has a $100 bill. He wants to buy three T-shirts for $18 each and tennis shoes for $41. Does he have enough money?

5. Garry drove to visit his friend Henry. How far apart do the friends live if Garry spent 3 hours driving and his speed was 68 miles per hour?

6. Linda is reading an interesting book. How long will she read it if the book has 448 pages and Linda reads 46 pages per hour?

7. Andrew has $ 1,299 on his checking account. He deposited $980, then $523, but then he withdrew $768. How much money does he have on his account after all these transactions?

8. A school has bought 9 new tables and 27 new chairs for the cafeteria. How much did the school pay if each table costs $73, and each chair costs $22?

9. A tennis club bought 22 new rackets. How much does each cost if the club paid $1,760 for this purchase?

10. Tom, a college student, drove 1,960 miles to visit his parents on Thanksgiving. How many hours did he drive if his average speed was 52 miles per hour?

Chapter 2 Test

EVALUATE

1. 25 - 5 x 3 =
2. 12 ÷ 4 + 4 x 2 =
3. 5 x (12 - 4) =
4. 20 ÷ 5 x (3 + 7) =
5. 32 ÷ (15 - 7) + 6 =
6. 4 x (3 + 2) x 5 =
7. Round to the nearest ten: 148, 83, 275
8. Round to the nearest thousand: 22,531, 167,289
9. Round to the nearest hundred million: 668,347,208
10. Round to the place of the underlined digit: 397,4$\underline{9}$8,611

ESTIMATE

11. Using rounding to the nearest hundred: 389 + 1,235 ≈
12. Using compensation: 99 + 37 ≈
13. Using rounding: 3,004 x 294 ≈

Word Problems
Solve by estimating.

14. During his college years, Michael visited his parents 8 times. His mother came to visit him 6 times, three times together with his father. How much money did the family spend for the tickets, if the average price for the round trip is $386?

15. The Simpsons went to a ski resort, which is 418 miles away from their house. What was their average speed if they reached the resort in 7 hours?

Chapter 3

Lesson 1 Homework 1

Is the first number a factor of the second? Answer yes or no.

1. 3, 18
2. 4, 14
3. 7, 28
4. 2, 130
5. 5, 125
6. 8, 36

Find all the factors of the given number.

7. 12
8. 18
9. 20
10. 28
11. 32
12. 36
13. 29
14. 45
15. 96

State the first five nonzero multiples of each number.

16. 4
17. 5
18. 9
19. 12
20. 20
21. 25

Is the statement true or false?

22. 6 is a multiple of 36.
23. 40 is a multiple of 8.
24. 24 is a factor of 6.
25. 7 is a factor of 56.
26. 56 is a multiple of 2.
27. 4 is a factor of 34.
28. 72 is a multiple of 9.

Lesson 1 Homework 2

Is the first number a factor of the second? Answer yes or no.

1. 4, 18
2. 6, 72
3. 9, 91
4. 7, 42
5. 11, 121
6. 12, 54

Find all the factors of the given number.

7. 15
8. 24
9. 40
10. 27
11. 31
12. 56
13. 60
14. 64
15. 72

State the first five nonzero multiples of each number.

16. 2
17. 6
18. 8
19. 13
20. 18
21. 30

Is the statement true or false?

22. 32 is a multiple of 8.
23. 7 is a multiple of 28.
24. 15 is a factor of 5.
25. 9 is a factor of 81.
26. 60 is a multiple of 10.
27. 6 is a factor of 46.
28. 144 is a multiple of 12.

Lesson 2 Homework 1

State whether or not the number is divisible by 2.

1. 8 2. 24 3. 157 4. 1036

State whether or not the number is divisible by 3.

5. 51 6. 28 7. 258 8. 192

State whether or not the number is divisible by 4.

9. 132 10. 324 11. 7,742 12. 23,924

State whether or not the number is divisible by 5.

13. 75 14. 280 15. 3,335 16. 873,421

State whether or not the number is divisible by 6.

17. 96 18. 471 19. 9,702 20. 725,364

State whether or not the number is divisible by 9.

21. 126 22. 738 23. 15,249 24. 467,325

State whether or not the number is divisible by 10.

25. 75 26. 320 27. 472,910 28. 96,241

Lesson 2 Homework 2

State whether or not the number is divisible by 2.

1. 84 2. 31 3. 378 4. 7,539

State whether or not the number is divisible by 3.

5. 201 6. 87 7. 4,321 8. 246,354

State whether or not the number is divisible by 4.

9. 712 10. 3,548 11. 55,336 12. 284,602

State whether or not the number is divisible by 5.

13. 375 14. 1,035 15. 5,553 16. 62,485

State whether or not the number is divisible by 6.

17. 106 18. 354 19. 27,648 20. 773,442

State whether or not the number is divisible by 9.

21. 477 22. 5,381 23. 246,357 24. 78,894

State whether or not the number is divisible by 10.

25. 720 26. 465 27. 83,990 28. 24,000

Lesson 3 Homework 1

Give all the factors of the number. State whether the number is prime or composite.

1. 18
2. 19
3. 32
4. 37
5. 48
6. 60

State whether each number is prime or composite. If it composite, give two factors of the number other than itself and 1.

7. 5
8. 8
9. 13
10. 16
11. 24
12. 27
13. 37
14. 49
15. 51
16. 57
17. 63
18. 67

19. Using the sieve of Eratosthenes find all the prime numbers in this list:

1	2	3	4	5	6	7	8	9	10	11	12
13	14	15	16	17	18	19	20	21	22	23	24
25	26	27	28	29	30	31	32	33	34	35	36
37	38	39	40	41	42	43	44	45	46	47	48
49	50	51	52	53	54	55	56	57	58	59	60
61	62	63	64	65	66	67	68	69	70	71	72
73	74	75	76	77	78	79	80	81	82	83	84
85	86	87	88	89	90	91	92	93	94	95	96
97	98	99	100								

Lesson 4 — Homework 1

Write each number as the product of two factors.

1. 12
2. 13
3. 25
4. 30
5. 42
6. 47

Complete each factor tree to find the prime factorization of each number.

7.
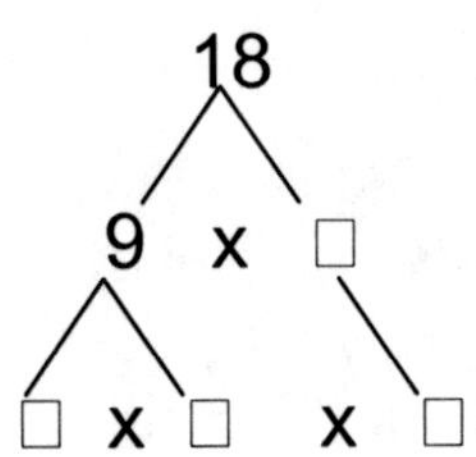

8.
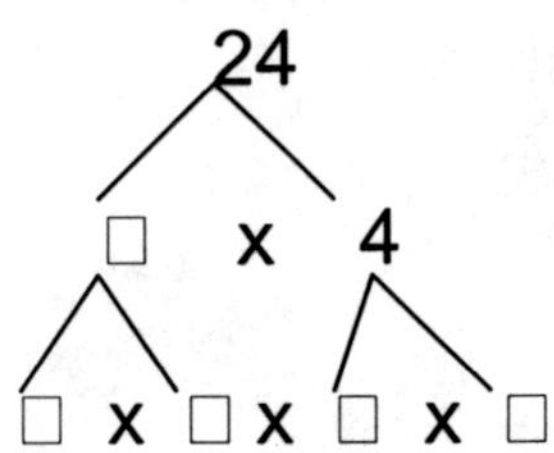

9.
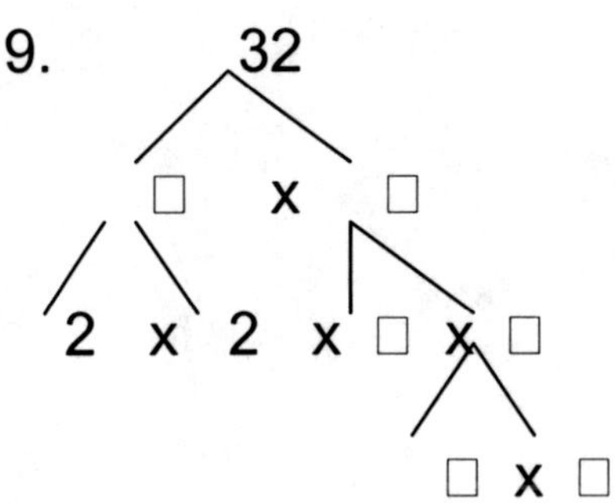

Make a factor tree to find the prime factorization of each number.

10. 36
11. 50
12. 78

Complete the "ladder" to find the prime factorization of each number.

13.
```
2 | 48
  □ | 24
    □ | 12
      □ | 6
        □ | 3
            | 1
```

14.
```
3 | 81
  □ | 27
    □ | 9
      □ | 3
          | 1
```

Use repeated division to find the prime factorization of each number.

15. 96
16. 144
17. 225
18. 98
19. 120
20. 250

Lesson 4 Homework 2

Write each number as the product of two factors.

1. 15
2. 24
3. 31
4. 35
5. 41
6. 56

Complete each factor tree to find the prime factorization of each number.

7. 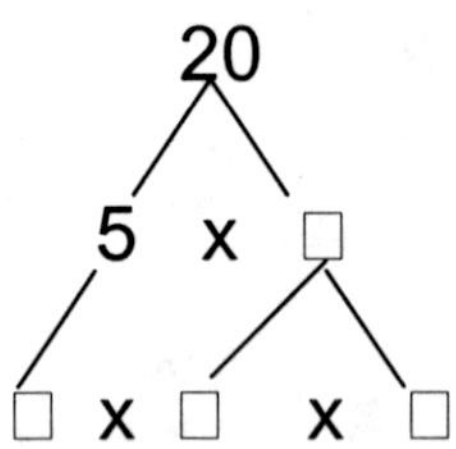

8. 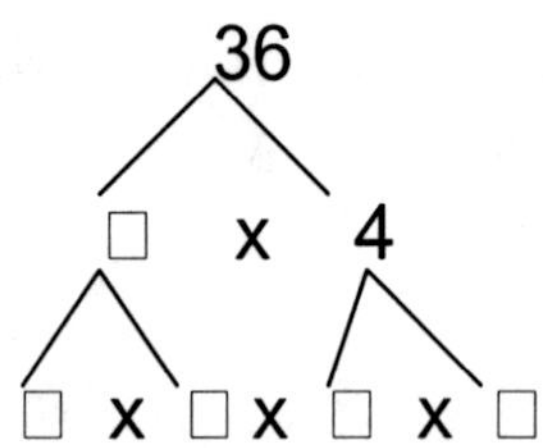

9. 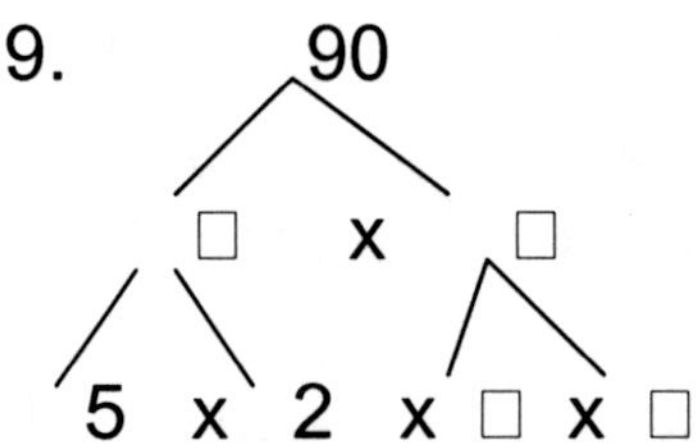

Make a factor tree to find the prime factorization of each number.

10. 42
11. 96
12. 72

Complete the "ladder" to find the prime factorization of each number.

13.
```
2 | 54
  □ | 27
    □ | 9
      □ | 3
          | 1
```

14.
```
2 | 250
  □ | 125
    □ | 25
      □ | 5
          | 1
```

Use repeated division to find the prime factorization of each number.

15. 48
16. 100
17. 108
18. 32
19. 54
20. 216

Lesson 5 Homework 1

Give the GCF of the two numbers. The set of factors of one number is given.

1. 12: [1, 2, 3, 4, 6, 12] , 18

2. 16: [1, 2, 4, 8, 16] , 24

3. 42: [1, 2, 3, 6, 7, 14, 21, 42] , 56

List the factors of both numbers to find the GCF of each pair.

4. 8, 10
5. 15, 20
6. 21, 28
7. 6, 14
8. 16, 40
9. 18, 24

List the factors of the smaller number to find the GCF of each pair.

10. 30, 45
11. 14, 64
12. 32, 48
13. 15, 65
14. 22, 121
15. 27, 72

Use prime factorization to find the GCF of each pair.

16. 24, 36
17. 50, 75
18. 42, 63
19. 54, 81
20. 25, 38
21. 90, 120

Use any method to find the GCF of each set of numbers.

22. 18, 24, 30
23. 14, 21, 28
24. 30, 45, 75
25. 32, 56, 72
26. 12, 20, 36
27. 22, 33, 55

Lesson 5 Homework 2

Give the GCF of the two numbers. The set of factors of one number is given.

1. 10: [1, 2, 5, 10] , 12

2. 18: [1, 2, 3, 6, 9, 18] , 24

3. 48: [1, 2, 3, 4, 6, 8, 12, 16, 24, 48] , 64

List factors of both numbers to find the GCF of each pair.

4. 4, 6
5. 12, 15
6. 16, 36
7. 9, 33
8. 20, 30
9. 27, 45

List the factors of the smaller number to find the GCF of each pair.

10. 28, 40
11. 22, 77
12. 24, 60
13. 13, 65
14. 42, 90
15. 27, 108

Use prime factorization to find the GCF of each pair.

16. 21, 28
17. 32, 56
18. 45, 105
19. 144, 108
20. 56, 168
21. 65, 91

Use any method to find the GCF of each set of numbers.

22. 9, 15, 21
23. 18, 27, 45
24. 25, 40, 55
25. 48, 56, 72
26. 33, 55, 121
27. 26, 39, 78

Lesson 6 Homework 1

Give five nonzero multiples of each number.

1. 7
2. 3
3. 13
4. 9

List multiples of both numbers to find the LCM of each pair.

5. 6, 8
6. 3, 5
7. 12, 15
8. 4, 10
9. 8, 24
10. 7, 8

List multiples of the larger number to find the LCM of each pair.

11. 7, 21
12. 15, 25
13. 8, 20
14. 6, 10
15. 9, 12
16. 8, 18

Use prime factorization to find the LCM of each pair.

17. 12, 15
18. 8, 18
19. 20, 25
20. 16, 24
21. 42, 70
22. 60, 72
23. 30, 36
24. 56, 63
25. 22, 55

Find the LCM of the numbers given using any method.

26. 18, 45
27. 72, 84
28. 70, 80
29. 3, 4, 8
30. 6, 15, 18
31. 30, 40, 60
32. 6, 10, 12
33. 9, 12, 36
34. 10, 15, 20

Lesson 6 Homework 2

Give five nonzero multiples of each number.

1. 5
2. 4
3. 10
4. 21

List multiples of both numbers to find the LCM of each pair.

5. 4, 6
6. 2, 7
7. 5, 9
8. 12, 15
9. 6, 8
10. 12, 18

List multiples of the larger number to find the LCM of each pair.

11. 9, 21
12. 4, 10
13. 12, 16
14. 16, 20
15. 27, 36
16. 18, 60

Use prime factorization to find the LCM of each pair.

17. 12, 16
18. 25, 30
19. 14, 21
20. 18, 21
21. 42, 48
22. 36, 60
23. 35, 56
24. 45, 60
25. 33, 44

Find the LCM of the numbers given using any method.

26. 11, 121
27. 63, 72
28. 35, 45
29. 4, 5, 6
30. 7, 14, 21
31. 12, 15, 20
32. 25, 30, 50
33. 12, 24, 36
34. 24, 32, 40

Lesson 7 Homework 1

Word problems.

1. Two college students, Henry and Tom, use the same laundromat. Henry goes to the laundromat every 8th day, and Tom goes every 6th day. If they both met in the laundromat on February 1, when will they meet there again?
2. Find two numbers whose product is 24 and whose LCM is 12.

3. Find two numbers whose product is 180 and whose LCM is 60.

4. Mrs. Bertolini, a seamstress, sews dresses for girls and women. Each girl's dress needs 3 meters of fabric; each woman's dress needs 4 meters. What is the minimum length of fabric that she should buy if she doesn't know whether she will be sewing girl's dresses or women's dresses?

5. Joan likes to read and goes to a library every 4th day. Her brother Edward doesn't like to read and only goes to the library every 9th day. If they both went to the library on March 1, when will they go there together again?

6. Jane is babysitting two babies. One eats every 3 hours and the second one eats every 4 hours. If Jane fed them together at 8 in the morning, when will she feed them together again?

7. For the first time in her life, Betty is home alone. Her mom calls her every 30 minutes and her dad calls her every 40 minutes. How often should Betty talk to both parents at the same time?

8. Julie is walking with her farther. Julie takes a step every 2 seconds. Her father takes a step every 3 seconds. How often do they take a step simultaneously?

9. Mrs. Jones has twin sons. During the night the first twin wakes up and screams every 30 minutes. The other twin wakes up screaming every 40 minutes. Between 8 p.m. and 8 a.m. how many times do they scream simultaneously?

10. Which is greater: the LCM of 6 and 8 or the LCM of 5 and 9?

Chapter 3 Test

Find all the factors of the given numbers.

1. 12 2. 32 3. 100

Find the first 5 nonzero multiples of the given numbers.

4. 7 5. 15 6. 21

Use a factor tree to find the prime factorization of each number.

7. 24 8. 60 9. 81

Use repeated division to find the prime factorization of each number.

10. 80 11. 144 12. 216

Find the GCF of each pair.

13. 18, 24 14. 42, 49 15. 56, 72

Find the LCM of each pair.

16. 24, 36 17. 12, 15 18. 25, 30

Word problems.

19. Find two numbers whose product is 36 and whose GCF is 3.

20. Find two numbers whose product is 96 and whose LCM is 24.

21. Mrs. Thompson goes to the grocery store every 4th day, while Mrs. Johnson goes to the same store every 7th day. If they met at the store on May 1, when will they meet there again?

Chapter 4

Lesson 1 Homework 1

Give the fraction of the region that is colored. Write the fraction as a number and with words.

1.

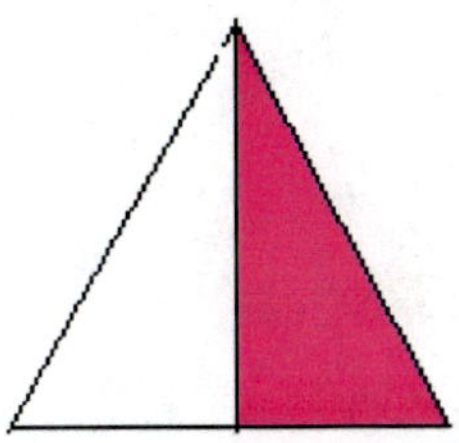

___, __________

2.

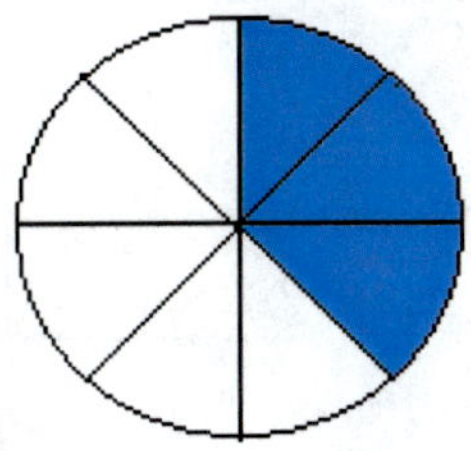

___, __________

3.

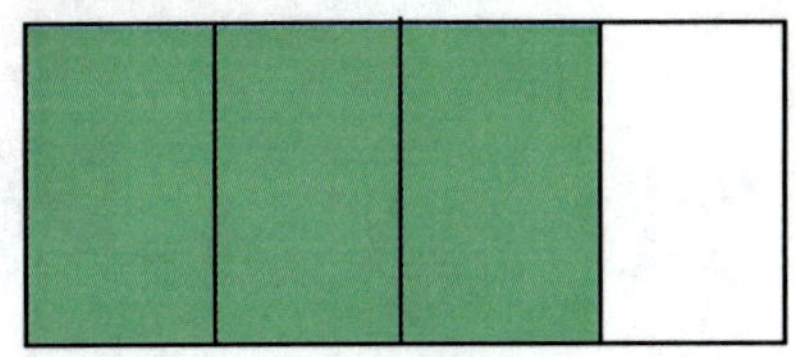

___, __________

4.

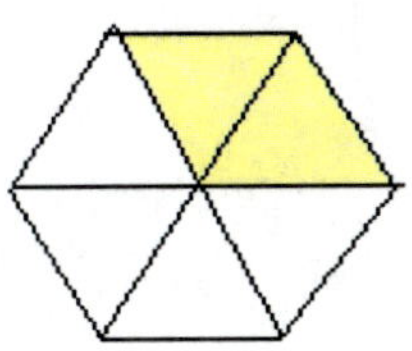

___, __________

5.

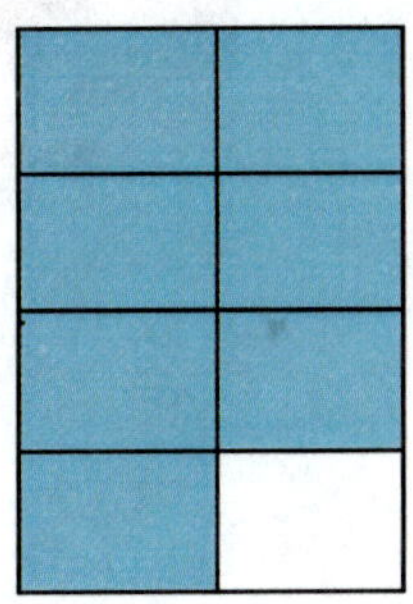

___, __________

6.

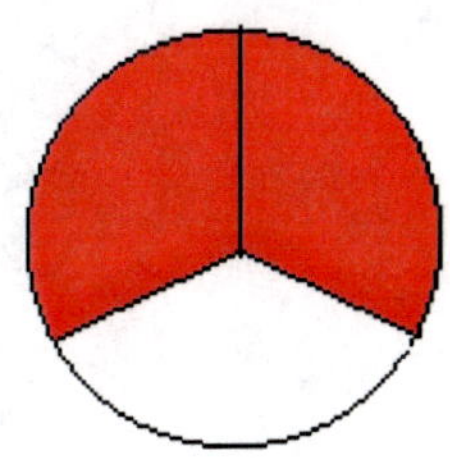

___, __________

Write the fractions.

7. one fourth = ____

8. two thirds = ____

9. four fifths = ____

10. seven eighths = ____

11. Show $\frac{3}{4}$ of the figure

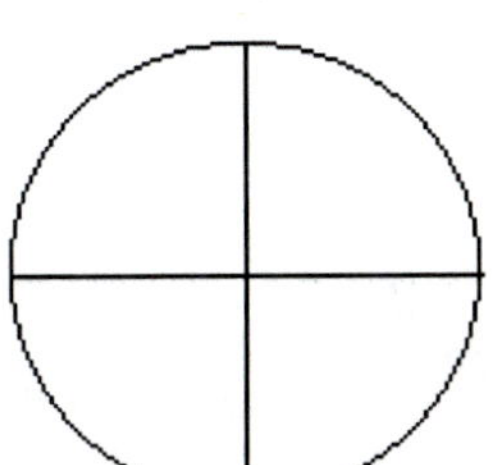

12. Show $\frac{7}{8}$ of the figure

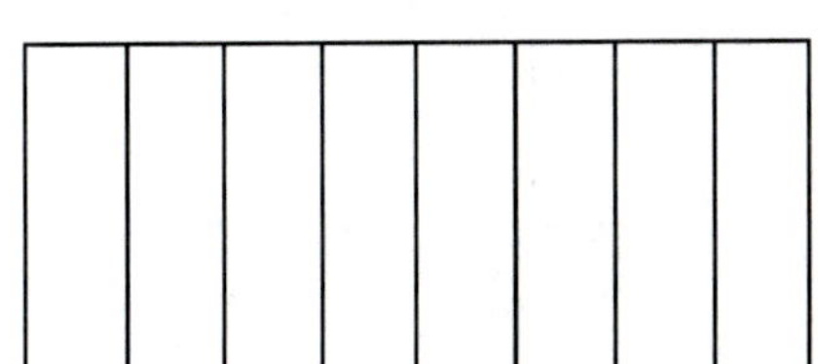

Lesson 1 Homework 2

Give the fraction of the region that is colored. Write the fraction as a number and with words.

1.

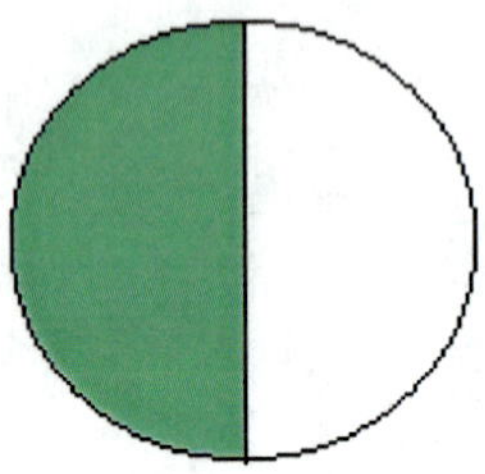

___, ________

2.

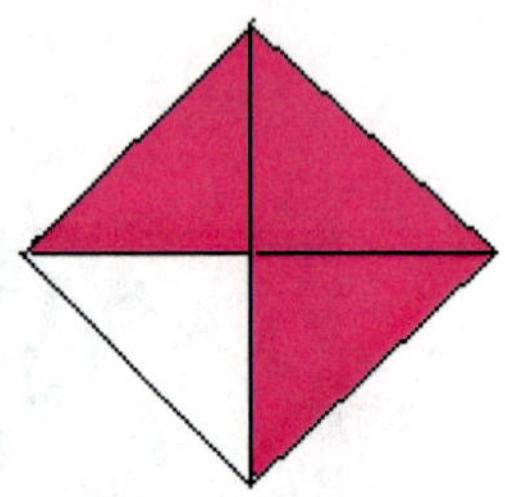

___, ________

3.

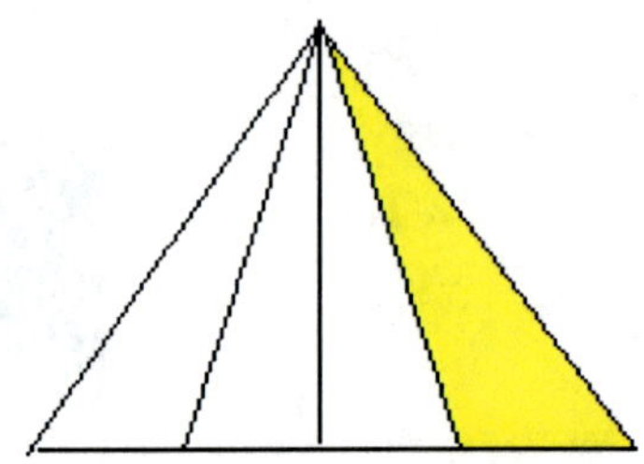

___, ________

4.

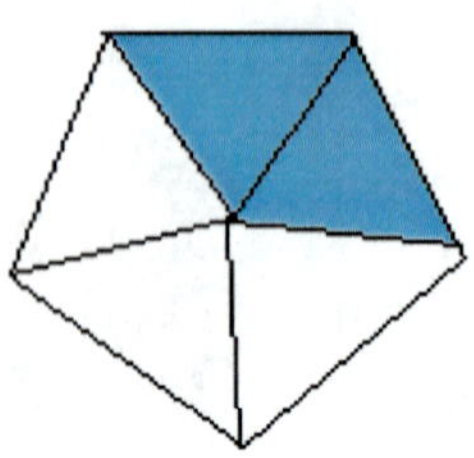

___, ________

5.

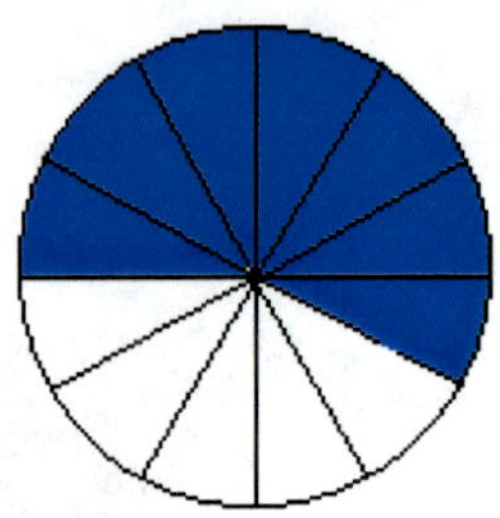

___, ________

6.

___, ________

Write the fractions.

7. three sevenths = ___

8. five tenths = ___

9. ten elevenths = ___

10. four thirteenths = ___

11. Show $\frac{4}{7}$ of the figure

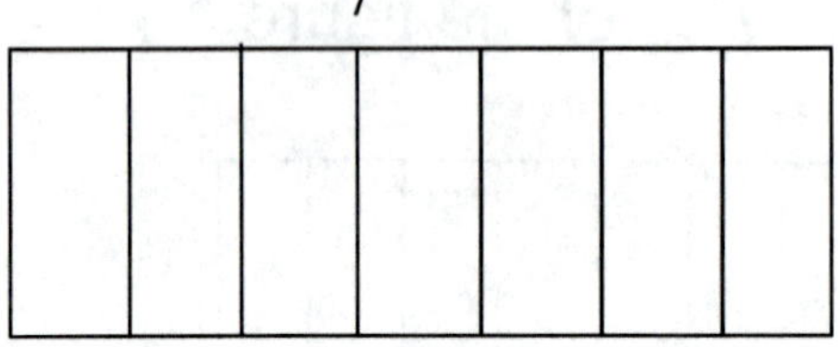

12. Show $\frac{2}{3}$ of the figure

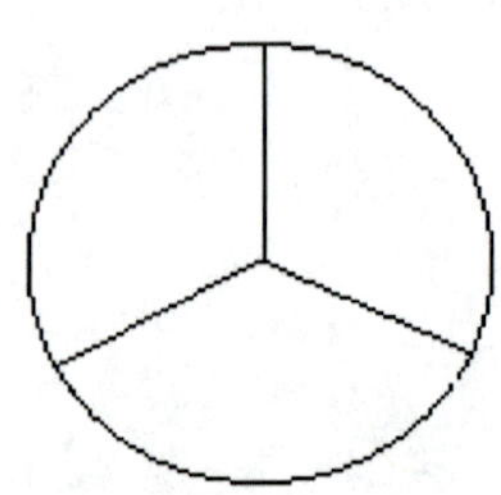

Lesson 2 Homework 1

What fraction stands for the colored part of each group?

1.

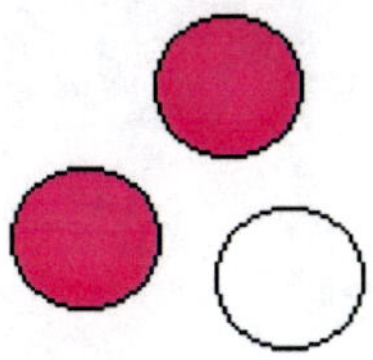

______

2.

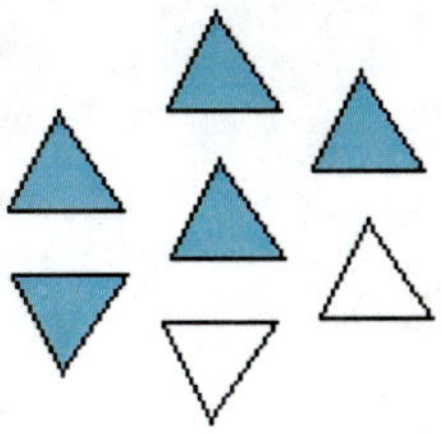

______

3.

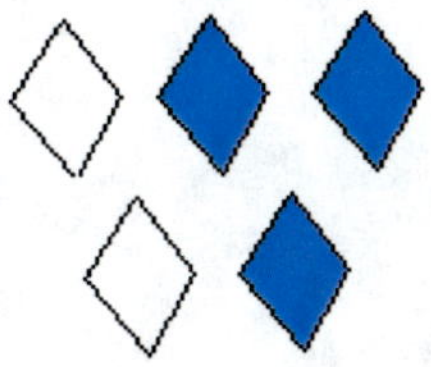

______

4.

______

5. There are 12 roses in a vase. Four of them are red, five are white, and three are yellow. Answer the questions:

1) What fractions of the flowers are red?

2) What fractions of the flowers are not white?

Write the fractions.

6. two sevenths = _____ 7. one half = _____

Write the word name for each fraction.

8. $\frac{7}{8} =$ 9. $\frac{3}{10} =$ 10. $\frac{6}{13} =$

Find each number.

11. $\frac{1}{5}$ of 35 12. $\frac{3}{4}$ of 16 13. $\frac{7}{10}$ of 50

Lesson 2 Homework 2

Give the fraction of each set that is colored.

1.

______

2.

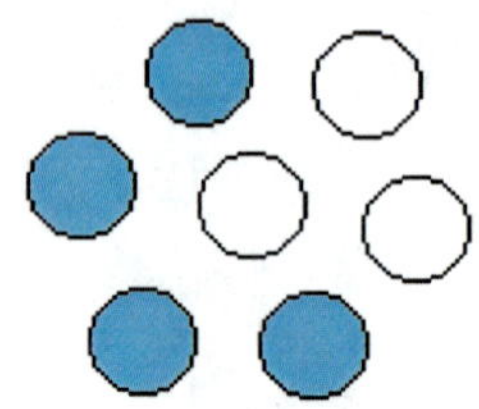

______

3.

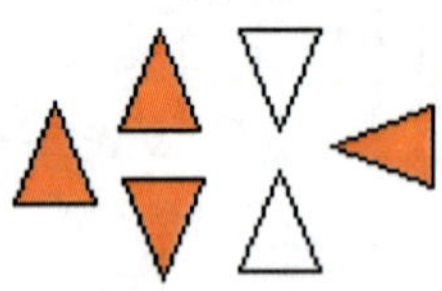

______

4.

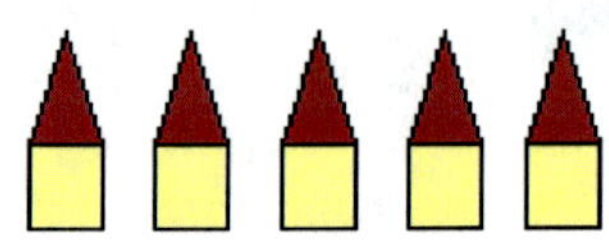

______

5. There are 15 people in a room. Six of them are boys, five are girls, two are women, and two are men. Answer the questions:

1) What fraction of the people are boys?

2) What fraction of the people are girls?

3) What fraction of the people are not kids ?

Show each fraction as part of a set.

6. $\frac{4}{5}$ 7. $\frac{3}{4}$ 8. $\frac{7}{10}$

Find each number.

9. $\frac{1}{8}$ of 24 10. $\frac{4}{5}$ of 40 11. $\frac{5}{11}$ of 33

Lesson 3 Homework 1

Write two fractions that name the colored part of each region or set.

1.

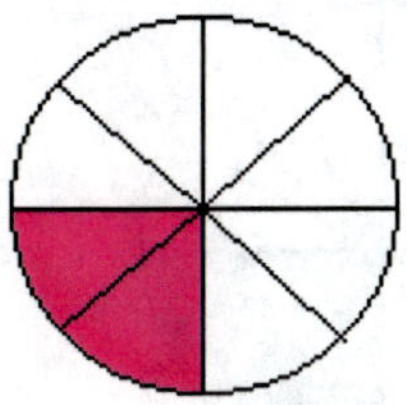

___ = ___

2.

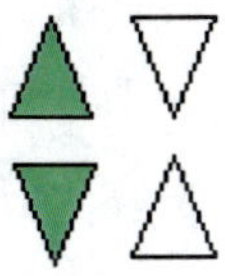

___ = ___

3.

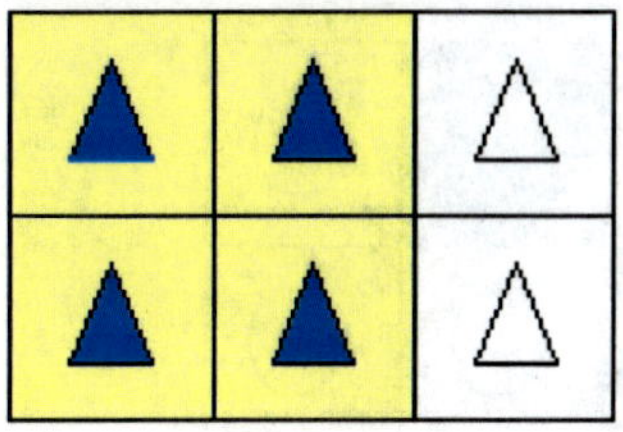

___ = ___

Write the equivalent fractions by multiplying the numerator and denominator by 2.

4. $\frac{2}{3} =$

5. $\frac{5}{8} =$

6. $\frac{7}{12} =$

Write the equivalent fractions by multiplying the numerator and denominator by 3.

7. $\frac{3}{4} =$

8. $\frac{4}{7} =$

9. $\frac{2}{5} =$

Give the next three equivalent fractions.

10. $\frac{3}{5} = - = - = -$

11. $\frac{2}{9} = - = - = -$

12. $\frac{4}{7} = - = - = -$

13. $\frac{5}{6} = - = - = -$

Lesson 3 — Homework 2

Write two fractions that name the colored part of each region or set.

1.

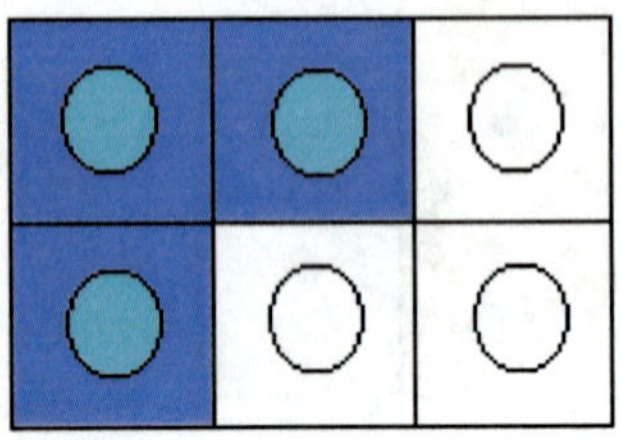

2.

3.

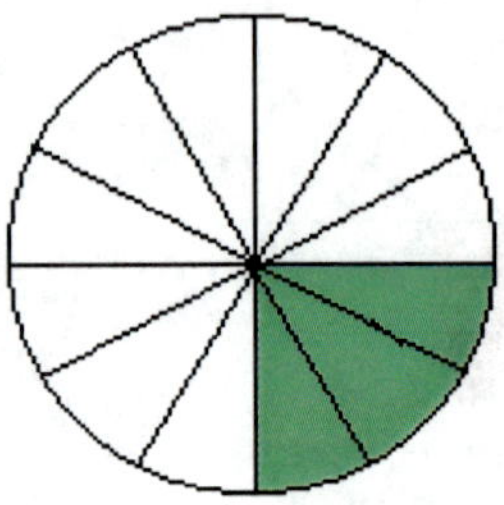

__ = __   __ = __   __ = __

Write the equivalent fractions by dividing the numerator and denominator by 2.

4. $\frac{2}{4} =$ 5. $\frac{6}{8} =$ 6. $\frac{8}{12} =$

Write the equivalent fractions by multiplying the numerator and denominator by 4.

7. $\frac{3}{4} =$ 8. $\frac{4}{7} =$ 9. $\frac{2}{5} =$

Give the next three equivalent fractions.

10. $\frac{1}{3} = \frac{}{} = \frac{}{} = \frac{}{}$ 11. $\frac{2}{5} = \frac{}{} = \frac{}{} = \frac{}{}$

12. $\frac{5}{7} = \frac{}{} = \frac{}{} = \frac{}{}$ 13. $\frac{11}{12} = \frac{}{} = \frac{}{} = \frac{}{}$

Lesson 3

Homework 3

Find the equivalent fractions. Write the number for each □.

1. $\frac{2}{3} = \frac{6}{\square}$
2. $\frac{3}{4} = \frac{\square}{8}$
3. $\frac{4}{5} = \frac{20}{\square}$
4. $\frac{7}{8} = \frac{\square}{24}$
5. $\frac{5}{9} = \frac{25}{\square}$
6. $\frac{1}{6} = \frac{\square}{24}$
7. $\frac{4}{7} = \frac{16}{\square}$
8. $\frac{2}{5} = \frac{\square}{50}$
9. $\frac{9}{11} = \frac{45}{\square}$
10. $\frac{11}{12} = \frac{\square}{72}$
11. $\frac{8}{15} = \frac{64}{\square}$
12. $\frac{10}{11} = \frac{\square}{99}$
13. $\frac{6}{7} = \frac{\square}{84}$
14. $\frac{7}{12} = \frac{49}{\square}$
15. $\frac{1}{8} = \frac{\square}{56}$
16. $\frac{10}{13} = \frac{80}{\square}$
17. $\frac{9}{14} = \frac{\square}{56}$
18. $\frac{25}{27} = \frac{50}{\square}$
19. $\frac{3}{10} = \frac{\square}{70}$
20. $\frac{30}{37} = \frac{90}{\square}$
21. $\frac{17}{20} = \frac{\square}{100}$

Lesson 4 Homework 1

Tell whether the fraction is in lowest terms.

1. $\frac{2}{3}$
2. $\frac{3}{6}$
3. $\frac{4}{12}$
4. $\frac{5}{8}$
5. $\frac{6}{9}$
6. $\frac{7}{10}$
7. $\frac{1}{6}$
8. $\frac{9}{18}$

Reduce to lowest terms.

9. $\frac{2}{8}$ = __
10. $\frac{4}{12}$ = __
11. $\frac{5}{15}$ = __
12. $\frac{6}{9}$ = __
13. $\frac{8}{10}$ = __
14. $\frac{9}{12}$ = __
15. $\frac{2}{6}$ = __
16. $\frac{3}{9}$ = __
17. $\frac{4}{16}$ = __
18. $\frac{5}{20}$ = __
19. $\frac{10}{30}$ = __
20. $\frac{7}{28}$ = __
21. $\frac{4}{20}$ = __
22. $\frac{12}{15}$ = __
23. $\frac{13}{26}$ = __
24. $\frac{14}{21}$ = __
25. $\frac{15}{18}$ = __
26. $\frac{16}{24}$ = __
27. $\frac{9}{45}$ = __
28. $\frac{11}{44}$ = __
29. $\frac{20}{100}$ = __

Lesson 4 Homework 2

Tell whether the fraction is in lowest terms.

1. $\frac{3}{9}$
2. $\frac{7}{8}$
3. $\frac{4}{18}$
4. $\frac{8}{21}$
5. $\frac{1}{4}$
6. $\frac{9}{15}$
7. $\frac{22}{25}$
8. $\frac{15}{18}$

Reduce to lowest terms when not already in lowest terms.

9. $\frac{5}{10}$ = __
10. $\frac{6}{14}$ = __
11. $\frac{8}{15}$ = __
12. $\frac{7}{12}$ = __
13. $\frac{9}{63}$ = __
14. $\frac{12}{30}$ = __
15. $\frac{10}{25}$ = __
16. $\frac{11}{33}$ = __
17. $\frac{25}{28}$ = __
18. $\frac{14}{18}$ = __
19. $\frac{16}{21}$ = __
20. $\frac{30}{40}$ = __
21. $\frac{15}{21}$ = __
22. $\frac{24}{28}$ = __
23. $\frac{18}{81}$ = __
24. $\frac{35}{42}$ = __
25. $\frac{54}{63}$ = __
26. $\frac{32}{45}$ = __

Lesson 4

Homework 3

Reduce to lowest terms if not already in lowest terms.

1. $\frac{4}{12}$ = __
2. $\frac{6}{36}$ = __
3. $\frac{1}{4}$ = __
4. $\frac{3}{27}$ = __
5. $\frac{2}{32}$ = __
6. $\frac{8}{28}$ = __
7. $\frac{7}{42}$ = __
8. $\frac{12}{48}$ = __
9. $\frac{20}{33}$ = __
10. $\frac{9}{16}$ = __
11. $\frac{16}{24}$ = __
12. $\frac{11}{77}$ = __
13. $\frac{44}{55}$ = __
14. $\frac{45}{63}$ = __
15. $\frac{28}{32}$ = __
16. $\frac{4}{100}$ = __
17. $\frac{8}{400}$ = ___
18. $\frac{10}{1000}$ = __
19. $\frac{5}{100}$ = __
20. $\frac{10}{80}$ = __
21. $\frac{20}{600}$ = __
22. $\frac{18}{90}$ = __
23. $\frac{45}{50}$ = __
24. $\frac{63}{72}$ = __
25. $\frac{11}{121}$ = __
26. $\frac{12}{144}$ = __
27. $\frac{89}{90}$ = __

Lesson 5 — Homework 1

Rewrite these mixed numbers as improper fractions.

1. $6\frac{1}{3} = -$
2. $5\frac{4}{5} = -$
3. $8\frac{2}{7} = -$
4. $7\frac{2}{5} = -$
5. $9\frac{5}{7} = -$
6. $4\frac{7}{9} = -$
7. $11\frac{3}{7} = -$
8. $13\frac{6}{7} = -$
9. $10\frac{9}{10} = -$
10. $20\frac{4}{5} = -$
11. $21\frac{2}{3} = -$
12. $18\frac{3}{4} = -$

Rewrite these improper fractions as mixed numbers in lowest terms.

13. $\frac{23}{7} =$
14. $\frac{30}{9} =$
15. $\frac{19}{3} =$
16. $\frac{40}{9} =$
17. $\frac{49}{8} =$
18. $\frac{72}{5} =$
19. $\frac{67}{7} =$
20. $\frac{55}{4} =$
21. $\frac{100}{11} =$
22. $\frac{88}{9} =$
23. $\frac{59}{7} =$
24. $\frac{95}{8} =$

Lesson 5 Homework 2

Rewrite these mixed numbers as improper fractions.

1. $3\frac{2}{3} = -$
2. $5\frac{7}{8} = -$
3. $9\frac{3}{4} = -$
4. $7\frac{4}{7} = -$
5. $4\frac{5}{6} = -$
6. $8\frac{2}{5} = -$
7. $10\frac{3}{5} = -$
8. $12\frac{1}{3} = -$
9. $11\frac{6}{7} = -$
10. $17\frac{1}{6} = -$
11. $20\frac{4}{5} = -$
12. $25\frac{3}{4} = -$

Rewrite these improper fractions as mixed numbers in lowest terms.

13. $\frac{15}{7} =$
14. $\frac{20}{9} =$
15. $\frac{24}{5} =$
16. $\frac{36}{5} =$
17. $\frac{45}{8} =$
18. $\frac{50}{7} =$
19. $\frac{48}{5} =$
20. $\frac{61}{10} =$
21. $\frac{29}{6} =$
22. $\frac{100}{13} =$
23. $\frac{81}{11} =$
24. $\frac{135}{12} =$

Lesson 6 Homework 1

Word problems.

1. There are 27 students in a class. One third of them did not do their math homework. How many students came home with a bad grade?

2. Two girls had 9 apples and shared them unfairly. One ate two thirds of the pile, but the other one ate what was left. How many apples did the unlucky girl eat?

3. Daniel is reading a book. He likes it so much that the first day he read three fourths of the book. The next day he read 25 pages and finished the book. How many pages did he read the first day?

4. Max and Sam came to visit their grandmother. Max ate $\frac{2}{7}$ of her blueberry pie, but Sam ate $\frac{3}{7}$. How much did each eat if the pie weighted 700 grams?

5. Uncle William decided to visit his nephew Charles. After driving one third of his way, his car broke down, and Charles came to rescue his uncle from the highway. How far apart do they live if Charles drove 230 miles to save his uncle?

6. The numerator of a fraction is 5 units less than the denominator. In lowest terms it is equal to one half. Find the fraction.

7. One evening Emmy was home alone. She spent $\frac{2}{3}$ of the time watching TV and then she read a book. How long did she read the book if she was alone for 3 hours?

8. Playing basketball, Ben made 15 baskets out of 20 shots. What fraction of his shots did he make? (Give the answer in lowest terms).

9. One day Liz spent 90 minutes practicing the piano. Her teacher wants her to practice one and a half hour each day. Did she practice enough that day?

10. There are 35 students in a class and $\frac{4}{5}$ of them plan to go to college. How many students are not planning to go to college?

Chapter 4 Test

Give the fraction of the region that is colored.

1.

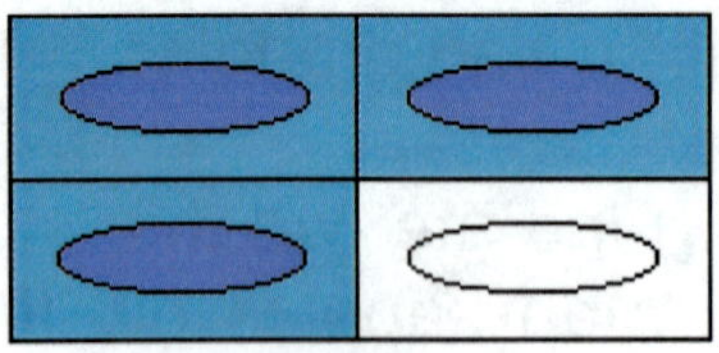

2.

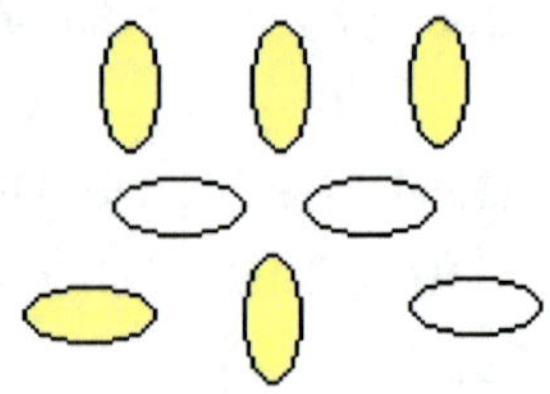

3. Show $\frac{7}{8}$ of the figure:

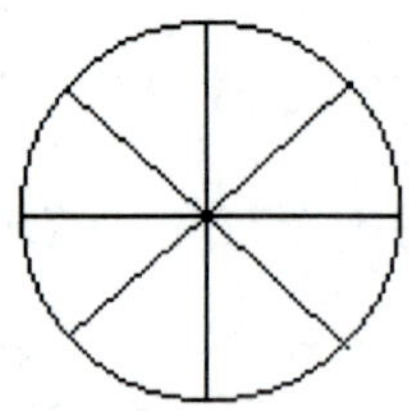

4. Michael has 10 balloons. Three of them are red, five are blue, and two are white. Answer the questions:

a) What fraction of the balloons is blue?

b) What fraction of the balloons is not red?

5. Find the equivalent fractions. Write the number for each $\square$.

a) $\frac{3}{4} = \frac{12}{\square}$ b) $\frac{5}{\square} = \frac{30}{48}$ c) $\frac{2}{7} = \frac{\square}{56}$

6. Reduce to lowest terms.

a) $\frac{3}{12} = -$ b) $\frac{16}{24} = -$ c) $\frac{14}{42} = -$

7. Write in order from greatest to least: $\frac{1}{3}, \frac{1}{2}, \frac{3}{8}, \frac{2}{5}, \frac{5}{8}$

8. Rewrite these mixed numbers as improper fractions.

a) $5\frac{3}{7} = -$ b) $9\frac{2}{5} = -$ c) $15\frac{1}{4} = -$

9. Solve: Amy gave a wonderful bouquet of flowers to her mom. $\frac{2}{3}$ of the flowers were roses; all the others were carnations. If there were 5 carnations, how many flowers were in the bouquet?

Chapter 5

Lesson 1 Homework 1

Add or subtract. Reduce if possible.

1. $\frac{1}{5}+\frac{3}{5}=$

2. $\frac{7}{8}-\frac{5}{8}=$

3. $\frac{3}{8}+\frac{1}{8}=$

4. $\frac{13}{20}-\frac{3}{20}=$

5. $\frac{4}{9}+\frac{2}{9}=$

6. $\frac{15}{16}-\frac{3}{16}=$

7. $\frac{8}{11}+\frac{3}{11}=$

8. $\frac{7}{11}-\frac{3}{11}=$

Add or subtract. Rename and reduce if possible.

9. $\frac{4}{7}+\frac{4}{7}=$

10. $\frac{7}{8}-\frac{5}{8}=$

11. $\frac{17}{20}+\frac{7}{20}=$

12. $\frac{13}{6}-\frac{5}{6}=$

13. $\frac{10}{13}+\frac{3}{13}=$

14. $\frac{8}{9}-\frac{2}{9}=$

15. $\frac{19}{20}+\frac{9}{20}=$

16. $\frac{25}{16}-\frac{7}{16}=$

Lesson 2 Homework 1

Add or subtract. Rename and reduce if possible.

1. $\frac{2}{3}+\frac{1}{9}=$

2. $\frac{1}{2}-\frac{1}{4}=$

3. $\frac{3}{10}+\frac{2}{5}=$

4. $\frac{5}{6}-\frac{1}{2}=$

5. $\frac{1}{3}+\frac{5}{12}=$

6. $\frac{1}{3}-\frac{1}{4}=$

7. $\frac{5}{6}+\frac{3}{8}=$

8. $\frac{3}{4}-\frac{1}{6}=$

9. $\frac{1}{2}+\frac{4}{5}=$

10. $\frac{7}{8}-\frac{5}{12}=$

11. $\frac{7}{15}+\frac{4}{5}=$

12. $\frac{2}{3}-\frac{3}{7}=$

13. $\frac{3}{4}+\frac{2}{9}=$

14. $\frac{5}{6}-\frac{7}{9}=$

15. $\frac{7}{10}+\frac{7}{15}=$

16. $\frac{9}{14}-\frac{8}{21}=$

17. $\frac{2}{3}+\frac{7}{12}=$

18. $\frac{8}{9}-\frac{5}{12}=$

19. $\frac{7}{8}+\frac{2}{5}=$

20. $\frac{5}{9}-\frac{1}{3}=$

Lesson 2

Homework 2

Add or subtract. Rename and reduce if possible.

1. $\frac{3}{4}+\frac{1}{2}=$

2. $\frac{1}{3}-\frac{1}{6}=$

3. $\frac{2}{7}+\frac{4}{21}=$

4. $\frac{1}{3}-\frac{1}{4}=$

5. $\frac{4}{5}+\frac{5}{6}=$

6. $\frac{7}{8}-\frac{1}{2}=$

7. $\frac{1}{6}+\frac{3}{8}=$

8. $\frac{11}{14}-\frac{5}{21}=$

9. $\frac{8}{9}+\frac{5}{6}=$

10. $\frac{11}{12}-\frac{3}{8}=$

11. $\frac{5}{8}+\frac{7}{10}=$

12. $\frac{7}{10}-\frac{4}{15}=$

13. $\frac{14}{15}+\frac{7}{20}=$

14. $\frac{25}{33}-\frac{9}{22}=$

15. $\frac{7}{12}+\frac{8}{9}=$

16. $\frac{3}{4}-\frac{7}{22}=$

17. $\frac{12}{13}+\frac{9}{26}=$

18. $\frac{7}{9}-\frac{3}{5}=$

19. $\frac{1}{3}+\frac{7}{8}=$

20. $\frac{5}{12}-\frac{1}{4}=$

Lesson 3 Homework 1

Add. Reduce to lowest terms.

1. $1+\frac{3}{7}=$

2. $1\frac{2}{7}+3\frac{3}{7}=$

3. $4\frac{9}{10}+6=$

4. $2\frac{1}{6}+5\frac{5}{6}=$

5. $3\frac{1}{3}+1\frac{1}{3}=$

6. $2+7\frac{7}{10}=$

7. $5+3\frac{6}{11}=$

8. $5\frac{7}{13}+4\frac{3}{13}=$

9. $6\frac{2}{7}+7\frac{4}{7}=$

10. $6\frac{7}{12}+1\frac{8}{12}=$

11. $8\frac{4}{9} + 11\frac{2}{9}$

12. $12\frac{9}{10} + 3\frac{3}{10}$

13. $12\frac{6}{7} + 8\frac{3}{7}$

14. $23\frac{8}{13} + 8\frac{7}{13}$

15. $8\frac{3}{7} + 7\frac{4}{7}$

16. $14\frac{5}{11} + 4\frac{8}{11}$

17. $6\frac{7}{10} + 13\frac{3}{10}$

18. $20\frac{2}{15} + 9\frac{8}{15}$

19. $\frac{8}{11} + 7\frac{3}{11}$

Lesson 3 Homework 2

Add or subtract. Reduce to lowest terms.

1. $2\frac{1}{5}+4\frac{3}{5}=$

2. $5\frac{3}{4}+2=$

3. $7\frac{3}{7}+6\frac{2}{7}=$

4. $8\frac{7}{8}+5\frac{3}{8}=$

5. $11\frac{7}{9}+8\frac{5}{9}=$

6. $23\frac{1}{3}+7\frac{2}{3}=$

7. $14\frac{3}{4}+16\frac{3}{4}=$

8. $27\frac{4}{7}+13\frac{5}{7}=$

9. $29\frac{8}{11}+32\frac{3}{11}=$

10. $33\frac{3}{8}+18\frac{1}{8}=$

11. $27\frac{7}{8} + 7\frac{7}{8}$

12. $41\frac{9}{14} + 13\frac{3}{14}$

13. $18\frac{11}{12} + 22\frac{7}{12}$

14. $50\frac{11}{14} + 27\frac{9}{14}$

15. $66\frac{8}{15} + 33\frac{7}{15}$

16. $63\frac{4}{7} + 23\frac{2}{7}$

17. $78\frac{9}{11} + 37\frac{7}{11}$

18. $125\frac{7}{13} + 38\frac{9}{13}$

19. $29\frac{5}{9} + 47\frac{7}{9}$

Lesson 4 Homework 1

Subtract. Rename and reduce if possible.

1. $4-\frac{2}{9}=$
2. $8-3\frac{1}{2}=$
3. $13\frac{4}{7}-8\frac{2}{7}=$
4. $41\frac{7}{10}-11\frac{3}{10}=$
5. $18\frac{7}{8}-12\frac{3}{8}=$
6. $35\frac{9}{11}-16\frac{2}{11}=$
7. $20\frac{1}{3}-10\frac{2}{3}=$
8. $7\frac{4}{7}-3\frac{6}{7}=$
9. $14\frac{2}{5}-13\frac{4}{5}=$
10. $8\frac{1}{9}-7\frac{4}{9}=$
11. $12\frac{3}{4} - 3\frac{1}{4}$
12. $8\frac{5}{6} - 6\frac{1}{6}$
13. $18\frac{7}{11} - 12\frac{2}{11}$
14. $43\frac{1}{4} - 8\frac{3}{4}$
15. $20\frac{5}{12} - 14\frac{7}{12}$
16. $60 - 8\frac{3}{4}$
17. $29 - 9\frac{1}{2}$
18. $31\frac{3}{10} - 16\frac{7}{10}$
19. $114\frac{2}{9} - 24\frac{5}{9}$

Lesson 4 Homework 2

Subtract. Rename and reduce if possible.

1. $7-\frac{3}{5}=$

2. $12-4\frac{4}{7}=$

3. $21\frac{5}{11}-5\frac{3}{11}=$

4. $38\frac{7}{8}-18\frac{5}{8}=$

5. $49\frac{5}{9}-12\frac{2}{9}=$

6. $50\frac{5}{12}-10\frac{1}{12}=$

7. $13\frac{1}{4}-7\frac{3}{4}=$

8. $29\frac{3}{8}-9\frac{5}{8}=$

9. $68\frac{1}{5}-22\frac{4}{5}=$

10. $32\frac{2}{7}-18\frac{5}{7}=$

11. $9\frac{5}{6} - 6\frac{1}{6}$

12. $20\frac{7}{8} - 15\frac{3}{8}$

13. $21\frac{7}{12} - 20\frac{5}{12}$

14. $17\frac{2}{9} - 9\frac{5}{9}$

15. $35\frac{1}{3} - 11\frac{2}{3}$

16. $24 - 14\frac{4}{13}$

17. $58\frac{3}{10} - 18\frac{9}{10}$

18. $100 - 29\frac{3}{8}$

19. $62\frac{2}{21} - 52\frac{5}{21}$

Lesson 5 Homework 1

Add or subtract. Reduce to lowest terms.

1. $3\frac{1}{2}+4\frac{1}{4}=$

2. $7\frac{3}{5}-3\frac{1}{2}=$

3. $6\frac{3}{8}+5\frac{1}{4}=$

4. $9\frac{1}{3}-2\frac{2}{9}=$

5. $10\frac{1}{3}+2\frac{2}{5}=$

6. $13\frac{5}{6}-8\frac{3}{5}=$

7. $12\frac{2}{7}+13\frac{9}{14}=$

8. $10\frac{7}{10}-5\frac{3}{5}=$

9. $5\frac{2}{3}+4\frac{5}{6}=$

10. $8\frac{1}{4}-3\frac{1}{2}=$

11. $8\frac{3}{4} + 7\frac{7}{8}$

12. $6\frac{2}{9} - 5\frac{2}{3}$

13. $9\frac{4}{5} + 11\frac{3}{4}$

14. $11 - 3\frac{2}{7}$

15. $10\frac{7}{10} + 20\frac{11}{15}$

16. $25 - 18\frac{9}{10}$

17. $14\frac{7}{9} + 6\frac{5}{6}$

18. $35 - \frac{11}{12}$

19. $29\frac{5}{7} + 21\frac{2}{3}$

Lesson 5 Homework 2

Add or subtract. Reduce to lowest terms.

1. $5\frac{2}{7}+7\frac{3}{14}=$

2. $7\frac{4}{5}-3\frac{3}{10}=$

3. $8\frac{3}{4}+4\frac{1}{6}=$

4. $12\frac{1}{2}-7\frac{1}{3}=$

5. $13\frac{7}{8}+8\frac{5}{6}=$

6. $15-6\frac{7}{13}=$

7. $21\frac{8}{9}+\frac{1}{6}=$

8. $12\frac{3}{7}-5\frac{4}{21}=$

9. $35\frac{7}{12}+15\frac{8}{15}=$

10. $28\frac{7}{8}-8\frac{5}{12}=$

11. $18\frac{8}{15} + 17\frac{7}{10}$

12. $23\frac{1}{4} - 20\frac{13}{14}$

13. $14\frac{13}{21} + 8\frac{11}{14}$

14. $40\frac{5}{16} - 15\frac{23}{24}$

15. $18\frac{4}{5} + 29\frac{5}{8}$

16. $80 - 39\frac{11}{12}$

17. $37\frac{7}{12} + 47\frac{7}{10}$

18. $62\frac{1}{9} - 61\frac{1}{6}$

19. $56\frac{3}{4} + 48\frac{7}{12}$

Lesson 5

Homework 3

Add or subtract. Reduce to lowest terms.

1. $3\frac{1}{2}+2\frac{2}{3}=$

2. $7\frac{3}{4}-4\frac{1}{2}=$

3. $5\frac{3}{4}+4\frac{5}{6}=$

4. $8\frac{1}{3}-6\frac{2}{5}=$

5. $9\frac{1}{9}+7\frac{1}{6}=$

6. $11\frac{3}{5}-9\frac{7}{10}=$

7. $10\frac{7}{8}+8\frac{3}{7}=$

8. $24\frac{11}{12}-4\frac{5}{8}=$

9. $17\frac{11}{12}+23\frac{8}{15}=$

10. $30-8\frac{15}{19}=$

11. $58\frac{17}{18}+12\frac{3}{4}=$

12. $45\frac{3}{4}-15\frac{19}{20}=$

13. $5\frac{1}{2}+7\frac{2}{3}+10\frac{3}{4}=$

14. $4\frac{3}{4}+8\frac{1}{6}+7\frac{7}{8}=$

15. $18\frac{7}{10}+12\frac{2}{3}+27\frac{3}{5}=$

16. $21\frac{3}{4}+12\frac{4}{15}+28\frac{7}{10}=$

17. $30-5\frac{1}{2}-7\frac{3}{4}=$

18. $12-6\frac{2}{3}-2\frac{1}{6}=$

19. $6\frac{5}{6}-3\frac{1}{5}-\frac{7}{15}=$

20. $25\frac{5}{8}-5\frac{3}{4}-5\frac{1}{3}=$

Lesson 6 Homework 1

Write >, <, or = for each □. Use equivalent fractions.

1. $\frac{1}{2}$ □ $\frac{3}{5}$
2. $\frac{3}{4}$ □ $\frac{6}{8}$
3. $\frac{4}{5}$ □ $\frac{5}{7}$
4. $\frac{3}{7}$ □ $\frac{8}{20}$
5. $\frac{7}{10}$ □ $\frac{9}{10}$
6. $\frac{9}{12}$ □ $\frac{6}{8}$
7. $\frac{1}{3}$ □ $\frac{1}{5}$
8. $\frac{7}{10}$ □ $\frac{7}{8}$
9. $\frac{5}{6}$ □ $\frac{5}{7}$

Write >, <, or = for each □. Use cross products.

10. $\frac{3}{8}$ □ $\frac{7}{19}$
11. $\frac{9}{27}$ □ $\frac{3}{9}$
12. $\frac{11}{12}$ □ $\frac{14}{15}$
13. $\frac{15}{17}$ □ $\frac{12}{15}$
14. $\frac{20}{21}$ □ $\frac{17}{18}$
15. $\frac{7}{9}$ □ $\frac{21}{27}$

Write the fractions in order from least to greatest.

16. $\frac{1}{4}$, $\frac{1}{6}$, $\frac{1}{3}$, $\frac{5}{12}$
17. $\frac{2}{3}$, $\frac{1}{2}$, $\frac{3}{5}$, $\frac{3}{4}$
18. $\frac{9}{10}$, $\frac{14}{17}$, $\frac{12}{19}$, $\frac{2}{3}$

Lesson 6 Homework 2

Write >, <, or = for each □. Use equivalent fractions.

1. $\frac{2}{3}$ □ $\frac{3}{4}$

2. $\frac{4}{7}$ □ $\frac{1}{2}$

3. $\frac{7}{8}$ □ $\frac{10}{11}$

4. $\frac{6}{11}$ □ $\frac{24}{44}$

5. $\frac{8}{11}$ □ $\frac{7}{10}$

6. $\frac{12}{15}$ □ $\frac{15}{18}$

7. $\frac{14}{15}$ □ $\frac{9}{10}$

8. $\frac{20}{40}$ □ $\frac{30}{60}$

9. $\frac{11}{13}$ □ $\frac{12}{13}$

Write >, <, or = for each □. Use cross products.

10. $\frac{9}{14}$ □ $\frac{8}{13}$

11. $\frac{22}{23}$ □ $\frac{23}{25}$

12. $\frac{16}{24}$ □ $\frac{12}{18}$

13. $\frac{11}{20}$ □ $\frac{15}{29}$

14. $\frac{18}{20}$ □ $\frac{10}{11}$

15. $\frac{25}{100}$ □ $\frac{27}{100}$

Write the fractions in order from greatest to least.

16. $\frac{1}{3}$, $\frac{7}{22}$, $\frac{4}{11}$, $\frac{3}{10}$

17. $\frac{9}{10}$, $\frac{14}{17}$, $\frac{12}{19}$, $\frac{2}{3}$

Lesson 7 Homework 1

Estimate each sum or difference.

1. $5\frac{7}{8}+7\frac{1}{20}\approx$

2. $12\frac{1}{18}-2\frac{6}{7}\approx$

3. $7\frac{1}{6}+8\frac{5}{12}+9\frac{9}{10}\approx$

4. $18\frac{11}{12}-6\frac{12}{13}\approx$

5. $3\frac{4}{9}+4\frac{3}{20}+2\frac{1}{12}\approx$

6. $52-37\frac{3}{16}\approx$

7. $16\frac{8}{15}+24\frac{9}{20}\approx$

8. $48\frac{8}{9}-18\frac{2}{21}\approx$

9. $25\frac{1}{5}+30\frac{2}{7}+15\frac{3}{8}\approx$

10. $120\frac{5}{13}-83\frac{7}{15}\approx$

11. $88\frac{3}{8}+111\frac{6}{11}\approx$

12. $69\frac{1}{20}-35\frac{13}{15}\approx$

13. $21\frac{9}{11}+32\frac{7}{9}+42\frac{1}{9}\approx$

14. $96\frac{20}{23}-16\frac{2}{21}\approx$

15. $17\frac{2}{17}+23\frac{1}{11}+30\frac{1}{12}\approx$

16. $77\frac{3}{35}-17\frac{2}{27}\approx$

17. $25\frac{7}{8}+38\frac{3}{7}+25\frac{2}{5}\approx$

18. $50\frac{13}{15}-20\frac{10}{11}\approx$

19. $9\frac{9}{10}+19\frac{19}{20}+29\frac{29}{30}\approx$

20. $100\frac{1}{35}-49\frac{49}{50}\approx$

Lesson 8 Homework 1

Word problems.

1. A rectangular lot has sides of $7\frac{1}{3}$ meters, and $12\frac{3}{4}$ meters. What is the perimeter of the lot?

2. Phil, a greedy boy, ate $\frac{7}{8}$ of a pie. His younger brother Bill ate what was left. What part of the pie did Bill eat?

3. Two friends, Joe and Josh, went fishing. Joe caught a catfish that weighed $1\frac{3}{5}$ pounds, and Josh caught a trout that weighed $2\frac{1}{2}$ pounds. What was the weight of the total catch?

4. Kathy ate $2\frac{3}{5}$ pounds of candies in an hour. Jim ate $3\frac{1}{3}$ pounds of candies in the same time. How many more candies did Jim eat than Kathy?

5. Mrs. Plumcott holding her daughter Marsha in her arms weighs $183\frac{2}{3}$ pounds. Without her daughter she weighs $157\frac{3}{4}$ pounds. Find Marsha's weight.

6. Jake spent $3\frac{1}{6}$ hours watching TV and then he spent only $\frac{2}{3}$ of an hour doing his homework. How much longer did he spend wasting his time than he did doing something useful?

7. Three girls, Ann, Gwen, and Doris, picked strawberries. Ann picked $4\frac{2}{7}$ pounds, Gwen - $5\frac{1}{3}$ pounds, and Doris - $2\frac{5}{7}$ pounds. How many pounds of strawberries did they have for strawberry jam?

8. The Appleby family bought a huge watermelon. The first day they ate $\frac{3}{5}$ of the watermelon, the next day - only $\frac{2}{9}$. After this nobody could even look at the watermelon and they invited their neighbor, who finished the watermelon. What part of the watermelon did the neighbor eat?

9. Robert had a wonderful vacation. He spent $\frac{1}{4}$ of his vacation visiting his parents in New Jersey. Then he visited his best friend in Florida and spent $\frac{2}{5}$ of his vacation there. After that he traveled abroad. What part of his vacation did he spend traveling abroad?

10. In Saint Paul, it rained for $\frac{1}{6}$ of the month of June. How many days were rainy?

Chapter 5 Test

Add or subtract. Reduce to lowest terms if possible.

1. $\frac{5}{12}+\frac{1}{3}=$

2. $\frac{5}{8}-\frac{1}{6}=$

3. $\frac{1}{10}+\frac{1}{15}=$

4. $\frac{7}{9}-\frac{7}{12}=$

5. $\frac{3}{8}+\frac{4}{5}=$

6. $\frac{14}{15}-\frac{7}{20}=$

7. $4+\frac{8}{13}=$

8. $2-\frac{3}{7}=$

9. $12-5\frac{1}{3}=$

10. $2\frac{1}{8}+5\frac{3}{4}=$

11. $6\frac{3}{5}-1\frac{1}{2}=$

12. $9\frac{5}{9}+4\frac{4}{9}=$

13. $11\frac{1}{2}-3\frac{2}{3}=$

14. $23\frac{7}{8}+7\frac{5}{6}=$

15. $42\frac{2}{7}-21\frac{7}{8}=$

Estimate.

16. $7\frac{7}{8}+9\frac{1}{21}\approx$

17. $18\frac{2}{19}-12\frac{11}{12}\approx$

18. $59\frac{15}{16}+41\frac{1}{9}\approx$

Word problems.

19. A chalkboard in a classroom is $3\frac{1}{3}$ meters long and $1\frac{1}{4}$ meters wide. Find the perimeter of the chalkboard.

20. Mr. Sims walks $2\frac{2}{5}$ miles in the morning and $1\frac{7}{9}$ miles in the evening. How much longer is his morning walk than his evening walk?

Chapter 6

Lesson 1 Homework 1

MULTIPLY

1. $12 \times \frac{3}{4} =$
2. $14 \times \frac{4}{7} =$
3. $\frac{5}{6} \times 18 =$
4. $\frac{5}{9} \times 27 =$
5. $45 \times \frac{7}{15} =$
6. $24 \times \frac{1}{6} =$
7. $\frac{9}{10} \times 40 =$
8. $\frac{11}{12} \times 12 =$
9. $48 \times \frac{7}{8} =$
10. $100 \times \frac{17}{20} =$
11. $\frac{8}{11} \times 55 =$
12. $\frac{14}{15} \times 30 =$
13. $12 \times \frac{3}{5} =$
14. $14 \times \frac{5}{9} =$
15. $\frac{6}{7} \times 15 =$
16. $\frac{8}{9} \times 10 =$
17. $21 \times \frac{4}{5} =$
18. $45 \times \frac{2}{11} =$
19. $\frac{5}{6} \times 48 =$
20. $\frac{3}{7} \times 35 =$

Lesson 2 Homework 1

Multiply. Reduce to lowest terms.

1. $\frac{3}{8} \times \frac{4}{9} =$

2. $\frac{5}{7} \times \frac{7}{8} =$

3. $\frac{4}{5} \times \frac{10}{11} =$

4. $\frac{6}{7} \times \frac{8}{9} =$

5. $\frac{7}{8} \times \frac{12}{35} =$

6. $\frac{9}{10} \times \frac{5}{6} =$

7. $\frac{5}{9} \times \frac{8}{9} =$

8. $\frac{14}{15} \times \frac{10}{21} =$

9. $\frac{13}{15} \times \frac{5}{7} =$

10. $\frac{16}{25} \times \frac{15}{28} =$

11. $\frac{12}{13} \times \frac{26}{27} =$

12. $\frac{3}{10} \times \frac{50}{57} =$

13. $\frac{18}{25} \times \frac{20}{21} =$

14. $\frac{21}{28} \times \frac{4}{7} =$

15. $\frac{9}{11} \times \frac{7}{10} =$

16. $18 \times \frac{5}{9} =$

17. $\frac{14}{15} \times \frac{8}{21} =$

18. $\frac{6}{7} \times 42 =$

Lesson 2

Homework 2

Multiply. Reduce to lowest terms.

1. $\frac{4}{5} \times \frac{5}{6} =$

2. $\frac{6}{11} \times \frac{7}{9} =$

3. $\frac{2}{3} \times \frac{1}{8} =$

4. $\frac{7}{12} \times \frac{8}{9} =$

5. $\frac{3}{4} \times \frac{5}{7} =$

6. $\frac{9}{10} \times \frac{7}{12} =$

7. $\frac{11}{26} \times \frac{13}{33} =$

8. $\frac{6}{7} \times \frac{3}{8} =$

9. $\frac{17}{18} \times \frac{6}{13} =$

10. $\frac{14}{15} \times \frac{9}{14} =$

11. $12 \times \frac{3}{8} =$

12. $\frac{7}{12} \times 16 =$

13. $\frac{1}{7} \times \frac{15}{16} =$

14. $\frac{8}{9} \times \frac{18}{23} =$

15. $\frac{10}{33} \times \frac{22}{25} =$

16. $\frac{16}{21} \times \frac{7}{24} =$

17. $\frac{6}{7} \times \frac{14}{27} =$

18. $\frac{25}{26} \times \frac{13}{15} =$

Lesson 3 Homework 1

Multiply. Reduce to lowest terms.

1. $7\frac{1}{2} \times 4\frac{2}{3} =$

2. $3\frac{3}{5} \times 3\frac{3}{4} =$

3. $2\frac{1}{6} \times 2\frac{4}{7} =$

4. $7\frac{1}{3} \times 5\frac{1}{4} =$

5. $4\frac{4}{5} \times 4\frac{3}{8} =$

6. $4\frac{2}{7} \times 8\frac{3}{4} =$

7. $3\frac{9}{10} \times 4\frac{4}{9} =$

8. $6\frac{3}{7} \times 8\frac{1}{6} =$

9. $9\frac{1}{3} \times 5\frac{5}{8} =$

10. $4\frac{1}{6} \times 4\frac{1}{5} =$

11. $13\frac{1}{2} \times 3\frac{1}{15} =$

12. $7\frac{7}{8} \times 8\frac{4}{9} =$

13. $1\frac{23}{25} \times 2\frac{3}{16} =$

14. $7\frac{5}{7} \times \frac{8}{9} =$

15. $6\frac{3}{10} \times 2\frac{1}{7} =$

16. $1\frac{17}{27} \times 5\frac{5}{8} =$

17. $5\frac{5}{9} \times 2\frac{5}{8} =$

18. $3\frac{9}{10} \times \frac{5}{26} =$

19. $11\frac{1}{5} \times 3\frac{1}{8} =$

20. $5\frac{1}{7} \times 2\frac{5}{8} =$

Lesson 3 Homework 2

Multiply. Reduce to lowest terms.

1. $7\frac{3}{4} \times 8 =$
2. $12 \times 2\frac{2}{3} =$
3. $18 \times 3\frac{1}{3} =$
4. $4\frac{2}{7} \times 21 =$
5. $4\frac{2}{3} \times 1\frac{1}{21} =$
6. $1\frac{2}{25} \times 1\frac{17}{18} =$
7. $10\frac{4}{5} \times 7\frac{7}{9} =$
8. $4\frac{1}{12} \times 2\frac{2}{21} =$
9. $5\frac{3}{7} \times 8\frac{3}{4} =$
10. $3\frac{1}{9} \times 5\frac{1}{4} =$
11. $6\frac{3}{8} \times 1\frac{11}{17} =$
12. $6\frac{2}{3} \times 7\frac{3}{4} =$
13. $36 \times 6\frac{5}{9} =$
14. $5\frac{2}{7} \times 28 =$
15. $8\frac{3}{4} \times 3\frac{9}{10} =$
16. $10\frac{1}{2} \times \frac{13}{14} =$
17. $6 \times 5\frac{4}{9} =$
18. $3\frac{3}{11} \times 6\frac{7}{8} =$
19. $6\frac{6}{11} \times 4\frac{8}{9} =$
20. $3\frac{1}{12} \times 16 =$

Lesson 4 Homework 1

Give the reciprocal of each number.

1. $\frac{1}{3}$
2. $\frac{4}{9}$
3. $\frac{7}{2}$
4. 5
5. $2\frac{3}{4}$
6. $3\frac{3}{5}$

Find the quotients. Check by multiplying.

7. $\frac{3}{4} \div \frac{5}{8} =$
8. $\frac{1}{3} \div \frac{1}{6} =$
9. $\frac{5}{7} \div \frac{10}{13} =$
10. $\frac{4}{5} \div \frac{6}{25} =$
11. $\frac{2}{5} \div \frac{7}{15} =$
12. $\frac{8}{11} \div \frac{12}{55} =$
13. $\frac{9}{10} \div \frac{6}{7} =$
14. $\frac{14}{15} \div \frac{21}{25} =$
15. $\frac{5}{9} \div \frac{5}{12} =$
16. $\frac{12}{13} \div \frac{15}{26} =$
17. $\frac{3}{20} \div \frac{18}{25} =$
18. $\frac{4}{7} \div \frac{16}{17} =$
19. $\frac{18}{25} \div \frac{6}{35} =$
20. $\frac{7}{8} \div \frac{21}{40} =$

Lesson 4

Homework 2

Give the reciprocal of each number.

1. $\frac{2}{5}$
2. $\frac{7}{3}$
3. $\frac{4}{11}$
4. 12
5. $5\frac{4}{7}$
6. $8\frac{1}{9}$

Find the quotient. Check by multiplying.

7. $\frac{5}{8} \div \frac{7}{12} =$
8. $\frac{3}{4} \div \frac{4}{3} =$
9. $\frac{1}{4} \div \frac{1}{12} =$
10. $\frac{7}{9} \div \frac{14}{15} =$
11. $\frac{8}{9} \div \frac{20}{21} =$
12. $\frac{11}{12} \div \frac{9}{16} =$
13. $\frac{3}{8} \div \frac{15}{16} =$
14. $\frac{21}{22} \div \frac{18}{55} =$
15. $\frac{7}{25} \div \frac{14}{25} =$
16. $\frac{16}{17} \div \frac{4}{17} =$
17. $\frac{24}{25} \div \frac{36}{55} =$
18. $\frac{30}{49} \div \frac{40}{77} =$
19. $\frac{51}{55} \div \frac{17}{33} =$
20. $\frac{25}{27} \div \frac{100}{111} =$

Lesson 5 Homework 1

Divide. Reduce to lowest terms. Check by multiplying.

1. $3\frac{1}{3} \div 6\frac{2}{3} =$
2. $7\frac{1}{2} \div 2\frac{1}{2} =$
3. $5 \div 3\frac{1}{3} =$
4. $6\frac{2}{3} \div 10 =$
5. $12\frac{3}{4} \div 2\frac{1}{8} =$
6. $10\frac{4}{5} \div 6\frac{3}{10} =$
7. $7\frac{3}{5} \div \frac{19}{20} =$
8. $\frac{6}{7} \div 5\frac{1}{7} =$
9. $12\frac{2}{7} \div 12\frac{2}{7} =$
10. $3\frac{3}{5} \div 8\frac{1}{10} =$
11. $7 \div \frac{14}{15} =$
12. $10\frac{2}{7} \div 1\frac{4}{5} =$
13. $9\frac{1}{3} \div 5\frac{4}{9} =$
14. $10\frac{5}{7} \div 3\frac{1}{8} =$
15. $17\frac{3}{5} \div 6\frac{2}{7} =$
16. $14\frac{2}{7} \div 3\frac{17}{21} =$
17. $9\frac{1}{11} \div 10 =$
18. $7\frac{5}{9} \div 5\frac{2}{3} =$
19. $17\frac{1}{7} \div 1\frac{19}{21} =$
20. $1\frac{8}{13} \div \frac{14}{65} =$

Lesson 5 Homework 2

Divide. Reduce to lowest terms. Check by multiplying.

1. $3\frac{4}{7} \div 2\frac{8}{21} =$

2. $12\frac{3}{5} \div 2\frac{4}{5} =$

3. $7\frac{7}{8} \div 7\frac{5}{7} =$

4. $9\frac{3}{5} \div 1\frac{11}{25} =$

5. $11 \div 18\frac{1}{3} =$

6. $9\frac{2}{7} \div 13 =$

7. $5\frac{13}{15} \div 13\frac{1}{5} =$

8. $14\frac{2}{7} \div 5\frac{5}{14} =$

9. $5\frac{1}{10} \div 1\frac{2}{15} =$

10. $7\frac{1}{14} \div 9\frac{3}{7} =$

11. $5\frac{2}{3} \div \frac{14}{15} =$

12. $14\frac{2}{3} \div 11 =$

13. $9\frac{1}{3} \div 4\frac{1}{2} =$

14. $25 \div 3\frac{3}{4} =$

15. $12\frac{1}{2} \div 3\frac{3}{4} =$

16. $12\frac{3}{4} \div 2\frac{1}{8} =$

17. $25\frac{5}{7} \div 7\frac{13}{21} =$

18. $8\frac{2}{11} \div 8\frac{4}{7} =$

19. $\frac{16}{11} \div 1\frac{1}{55} =$

20. $17\frac{1}{7} \div 3\frac{17}{21} =$

Lesson 6 Homework 1

Estimate the result of multiplication.

1. $3\frac{3}{4} \times 5\frac{1}{10} \approx$

2. $7\frac{1}{8} \times 7\frac{7}{12} \approx$

3. $9\frac{2}{3} \times 10\frac{1}{7} \approx$

4. $6\frac{4}{5} \times 8\frac{2}{15} \approx$

5. $12\frac{1}{11} \times 3\frac{5}{6} \approx$

6. $19\frac{6}{7} \times 2\frac{7}{8} \approx$

7. $7\frac{9}{10} \times 11\frac{1}{11} \approx$

8. $5\frac{5}{9} \times 6\frac{6}{13} \approx$

9. $10\frac{6}{7} \times 12\frac{1}{20} \approx$

10. $20\frac{1}{15} \times 3\frac{9}{10} \approx$

11. $8\frac{3}{20} \times 4\frac{8}{9} \approx$

12. $24\frac{13}{15} \times 4\frac{2}{15} \approx$

13. $15\frac{1}{8} \times 5\frac{2}{19} \approx$

14. $9\frac{19}{21} \times 6\frac{9}{10} \approx$

15. $24\frac{2}{17} \times 5\frac{2}{19} \approx$

16. $17\frac{3}{7} \times 2\frac{1}{30} \approx$

17. $29\frac{1}{3} \times 3\frac{4}{15} \approx$

18. $10\frac{2}{25} \times 10\frac{3}{25} \approx$

Lesson 6

Homework 2

Estimate the result of division.

1. $9\frac{1}{9} \div 2\frac{8}{11} \approx$
2. $7\frac{2}{15} \div 6\frac{9}{10} \approx$
3. $12\frac{2}{15} \div 4\frac{1}{12} \approx$
4. $18\frac{3}{10} \div 6\frac{2}{17} \approx$
5. $24\frac{8}{9} \div 5\frac{2}{13} \approx$
6. $30\frac{1}{7} \div 5\frac{11}{12} \approx$
7. $50 \div 7\frac{1}{7} \approx$
8. $40\frac{4}{15} \div 8 \approx$
9. $63\frac{2}{5} \div 8\frac{8}{9} \approx$
10. $46\frac{1}{12} \div 14\frac{12}{13} \approx$
11. $69\frac{7}{8} \div 10\frac{2}{17} \approx$
12. $100\frac{1}{3} \div 10\frac{1}{5} \approx$
13. $54\frac{3}{4} \div 5\frac{8}{45} \approx$
14. $88\frac{2}{21} \div 10\frac{33}{34} \approx$
15. $16\frac{2}{9} \div 4\frac{3}{16} \approx$
16. $400\frac{3}{7} \div 20\frac{4}{19} \approx$
17. $44\frac{7}{8} \div 15\frac{1}{15} \approx$
18. $33\frac{3}{10} \div 32\frac{5}{12} \approx$

Lesson 7 Homework 1

Solve word problems.

1. Mrs. Taylor bought $6\frac{3}{4}$ pounds of candies for her two daughters, Dorothy and Rita. In how many days will she have to buy candies again if Dorothy eats $\frac{1}{2}$ of a pound each day, and Rita eats $\frac{5}{8}$ of a pound each day?

2. Betty practices the piano $10\frac{1}{2}$ hours in a week. How long does she practice each day if she practices the same amount of time each day?

3. Two friends, Robert and Earl, decided to meet each other. They started to ride their bicycles from their houses at the same time and met each other in $2\frac{5}{8}$ hours. How far apart do the friends live if Robert rode his bicycle with the speed of $3\frac{3}{7}$ miles per hour, and Earl - $5\frac{1}{3}$ miles per hour?

4. The Carter family rented a cabin on a lake for $2\frac{1}{2}$ weeks and paid $200 for this wonderful vacation. What is the rent for one week?

5. When David became 16 years old he started to grow very fast. His parents noticed that he grew $4\frac{2}{3}$ inches in 7 months. How much did he grow each month on average?

6. Find the area of a triangle if its base is $7\frac{2}{7}$ meters and its height is $9\frac{1}{3}$ meters. (Hint: the area of a triangle is $\frac{1}{2}$ x base x height).

7. Mary works as a babysitter 6 days a week, $7\frac{1}{3}$ hours each day. What is her salary for one week if she gets $6\frac{1}{4}$ dollars per hour?

8. A third grade class bought $19\frac{1}{2}$ pounds of watermelons for their field trip. How many watermelons did the class buy if each of them weighs $3\frac{1}{4}$ of a pound?

9. Debby can read 9 pages per hour in English and only 3 pages per hour in French. How long will it take her to read one book in English and one book in French if each book has 62 pages?

10. A rectangular lot is $6\frac{1}{4}$ meters long and $2\frac{1}{15}$ meters wide. What is the area of the lot? (Hint: the area of a rectangle = length x width).

Chapter 6 Test

Multiply or divide. Reduce to lowest terms.

1. $\frac{3}{10} \times \frac{5}{12} =$
2. $\frac{7}{8} \times 24 =$
3. $30 \times \frac{2}{15} =$
4. $\frac{2}{3} \times \frac{9}{10} =$
5. $\frac{6}{35} \times \frac{7}{8} =$
6. $3\frac{1}{2} \times \frac{5}{7} =$
7. $4\frac{1}{2} \times 4\frac{2}{3} =$
8. $2\frac{4}{7} \times 6\frac{2}{9} =$
9. $9\frac{1}{11} \times 12\frac{1}{10} =$
10. $\frac{2}{3} \div \frac{8}{9} =$
11. $\frac{6}{11} \div \frac{9}{22} =$
12. $12 \div \frac{6}{7} =$
13. $\frac{5}{7} \div 15 =$
14. $\frac{21}{32} \div \frac{15}{16} =$
15. $3\frac{1}{3} \div 6\frac{2}{3} =$
16. $14\frac{2}{5} \div 8\frac{1}{10} =$
17. $9\frac{1}{3} \div 8\frac{2}{5} =$
18. $14\frac{2}{3} \div 6\frac{1}{9} =$

Word problems

19. A picture frame is $10\frac{5}{7}$ inches long and $9\frac{4}{5}$ inches wide. What is the area of the frame?
20. Mrs. Thompson bought $3\frac{3}{4}$ pounds of candies and divided them fairly among her 5 kids. How many pounds of candies did each child get?

Chapter 7

Lesson 1 Homework 1

Write in words.

1. 0.3
2. 0.07
3. 0.005
4. 0.25
5. 0.038
6. 0.248
7. 1.4
8. 28.05
9. 10.604

Write in expanded form.

10. 0.43
11. 3.207
12. 24.3
13. 321.123
14. 65.007
15. 0.879

Write as a decimal.

16. $\frac{7}{10}$
17. $\frac{52}{100}$
18. $\frac{382}{1000}$
19. $5\frac{3}{100}$
20. $12\frac{23}{1000}$
21. $3\frac{6}{10}$

Write the decimals.

22. eight tenths
23. two and forty seven thousandths
24. one hundred fifty-three thousandths
25. seventy five and ninety-four hundredths

Use place value to tell what each underlined digit means.

26. $3.\underline{4}56$
27. $0.30\underline{5}$
28. $34.5\underline{2}2$

Write as fractions

29. 0.3
30. 0.84
31. 2.72
32. 17.245

Lesson 1 Homework 2

Write as a decimal.

1. $\frac{9}{10}$
2. $\frac{37}{100}$
3. $\frac{457}{1000}$
4. $7\frac{3}{10}$
5. $28\frac{38}{1000}$
6. $45\frac{47}{100}$

Write in words and in expanded form.

7. 3.5
8. 0.05
9. 0.453
10. 0.204
11. 22.22
12. 17. 032

Write the decimals.

13. seven hundredths
14. twenty and three hundred sixty-nine thousandths
15. eight and eight tenths

Use place value to tell what each underlined digit means.

16. 0.2<u>3</u>
17. 4<u>5</u>.205
18. 6.34<u>9</u>

Write all the numbers from the box where:

7.293	23.928
14.328	6.097
5.997	44.729

19. the 9 has the value of 0.9
20. the 7 has the value of 0.007
21. the 2 has the value of 0.02

Write as fractions

22. 0.5
23. 3.22
24. 25.075
25. 0.349

Lesson 2 Homework 1

Write as a decimal.

1. $\frac{79}{10000}$ 2. $\frac{2345}{100000}$ 3. $\frac{877651}{1000000}$

4. $12\frac{3}{10000}$ 5. $5\frac{56789}{100000}$ 6. $18\frac{45007}{1000000}$

Write in words and in expanded form.

7. 0.00007 8. 2.123456 9. 9.004505

10. 34.000123 11. 0.25257 12. 12.3456

Write the decimals.

13. forty-five ten-thousandths
14. thirty-seven and four hundred sixty-three thousand two hundred seven millionths
15. two and three thousand five hundred sixty-nine hundred-thousandths

Use place value to tell what each underlined digit means.

16. $23.35790\underline{2}$ 17. $7.000\underline{3}2$

18. $\underline{4}57.09945$ 19. $0.624\underline{6}81$

Write all the numbers from the box where:

3.46015	0.023983
0.000153	82.22212
6.23101	0.35113

20. the 3 has a value of 0.000003

21. the 5 has a value of 0.00005

22. the 1 has a value of 0.0001

Write as fractions. Reduce if possible

23. 0.0045 24. 0.02555 25. 20.00032

Lesson 2 Homework 2

Write as a decimal.

1. $\frac{7834}{100000}$
2. $\frac{305}{10000}$
3. $\frac{234567}{1000000}$
4. $6\frac{59}{10000}$
5. $27\frac{7}{100000}$
6. $3\frac{357}{1000000}$

Write in words and in expanded form.

7. 3.00345
8. 27.205607
9. 0.03407
10. 0.999999
11. 635.6805
12. 4.000432

Write the decimals.

13. thirty-five thousand seven hundred seventy millionths
14. three and nine hundred-thousandths
15. seventy-five and three thousand ninety-six ten-thousandths

Use place value to tell what each underlined digit means.

16. $2.005\underline{6}7$
17. $0.23456\underline{7}$
18. $99.40\underline{4}04$
19. $0.0000\underline{5}5$

Write all the numbers from the box where:

0.234626	5.023456
21.22942	0.00312
7.020426	1.11624

20. the 2 has a value of 0.00002
21. the 4 has a value of 0.0004
22. the 6 has a value of 0.000006

Write as fractions. Reduce if possible

23. 0.0052
24. 24.00275
25. 3.072

Lesson 3 Homework 1

Give the decimal name for point A.

1.

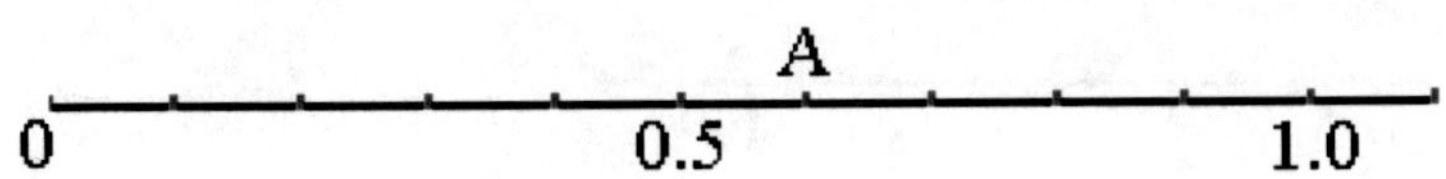

2.

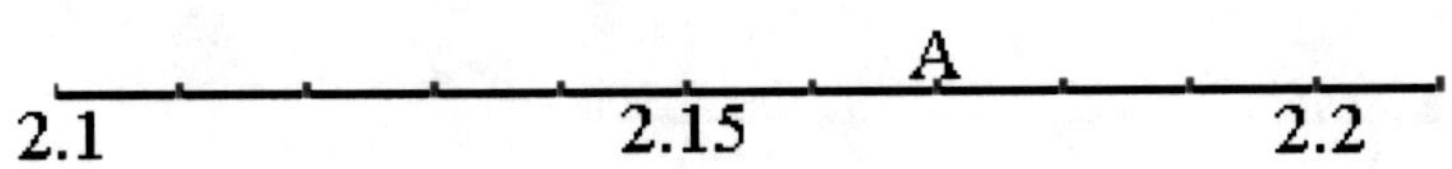

3.

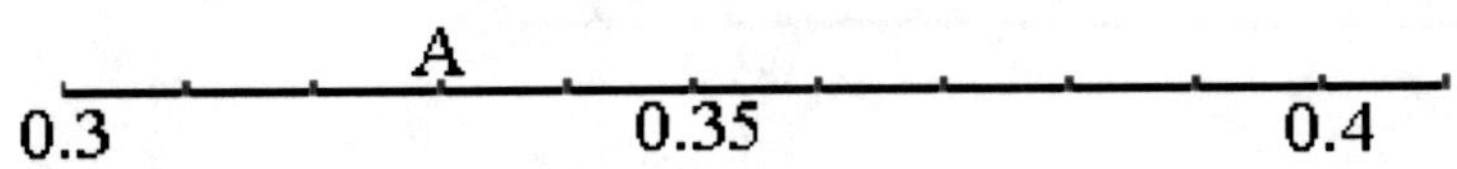

Write >, <, or = for each □.

4. 4.71 □ 4.709

5. 0.099 □ 0.12

6. 0.06453 □ 0.06543

7. 32.12 □ 32.1200

Write the numbers in order from least to greatest.

8. 0.0345, 0.0354, 0.0344

9. 2.17869, 2.17809, 2.17896, 2.17888

Give the decimal names for points A and B.

10.

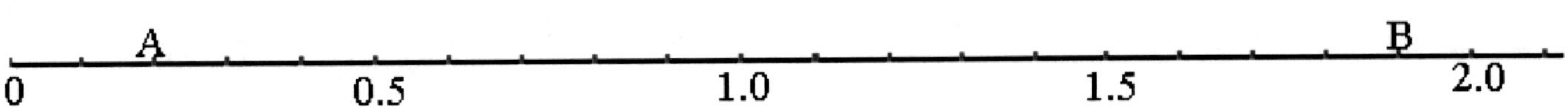

Make a number line and show each decimal on the number line.

11. 0.3

12. 1.25

13. 0.85

Lesson 3 — Homework 2

Give the decimal name for point P.

1.

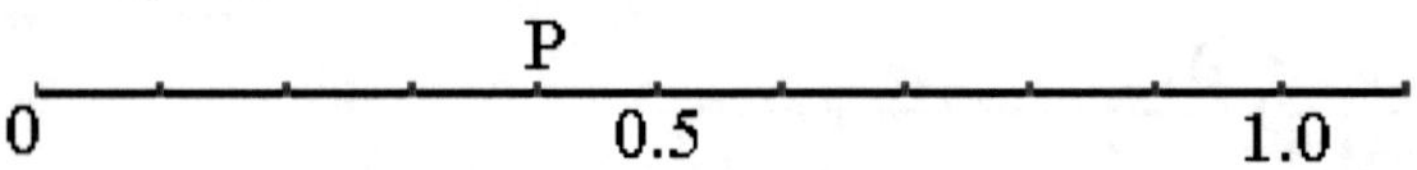

2.

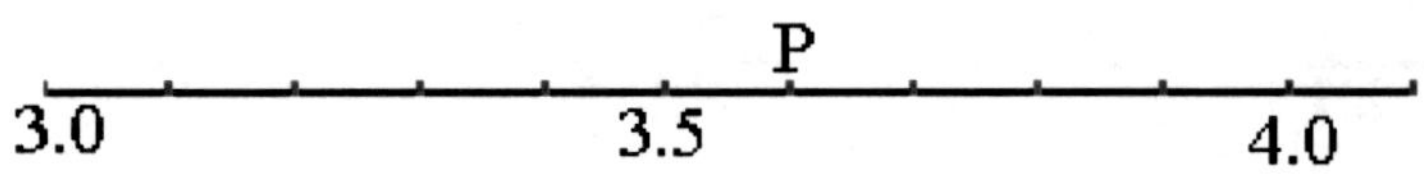

3.

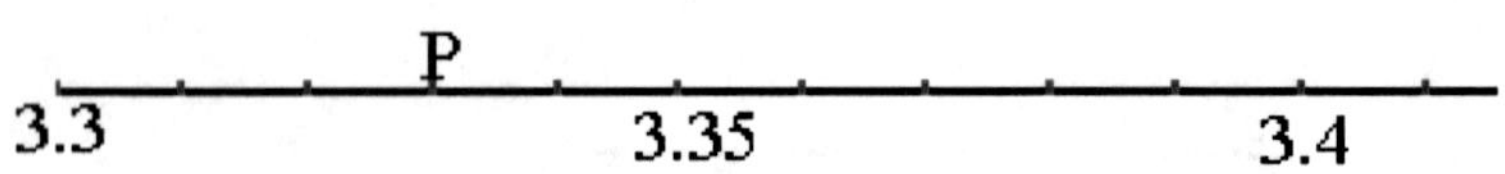

Write >, <, or = for each □.

4. 0.11 □ 0.109

5. 32.434 □ 32.44

6. 54.578 □ 54.587

7. 0.0305 □ 0.035

Write the numbers in order from greatest to least.

8. 7.8, 0.79, 8.0, 0.87

9. 0.0012, 0.0021, 0.0101, 0.00098

10. Give the decimal names for points A and B.

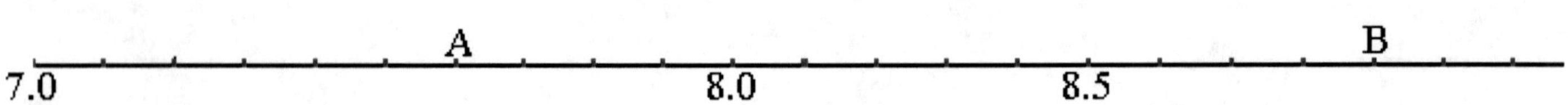

Make a number line and show each decimal on the number line.

11. 3.2

12. 3.45

13. 4.37

Lesson 4 — Homework 1

Round to the nearest whole number.

1. 0.564
2. 34.34
3. 4.8021

Round to the nearest tenth.

4. 7.649
5. 0.0623
6. 200.049

Round to the nearest hundredth.

7. 0.2345
8. 4.2253
9. 0.298

Round to the nearest thousandth.

10. 12.3334
11. 0.00452
12. 6.6688

Round to the nearest ten-thousandth.

13. 0.00345
14. 2.93554
15. 22.09996

Round to the place of the underlined digit.

16. $5.\underline{8}4$
17. $0.340\underline{2}8$
18. $0.002\underline{4}543$
19. $0.2\underline{9}9$
20. $2\underline{4}.582$
21. $0.056\underline{7}82$

Estimate using rounding (round to the nearest whole number).

22. $23.62 + 26.36 \approx$
23. $3.464 - 0.721 \approx$
24. $162.44 + 89.73 \approx$
25. $100 - 68.327 \approx$

Lesson 4 Homework 2

Round to the nearest whole number.

1. 0.329 2. 5.672 3. 29.666

Round to the nearest tenth.

4. 0.845 5. 32.052 6. 20.2099

Round to the nearest hundredth.

7. 0.0228 8. 26.2678 9. 6.44496

Round to the nearest thousandth.

10. 1.00678 11. 0.33333 12. 8.9999

Round to the nearest ten-thousandth.

13. 0.123456 14. 0.000086 15. 2.211221

Round to the place of the underlined digit.

16. $35.\underline{2}2$ 17. $6.03\underline{2}7$ 18. $0.0\underline{3}921$

19. $\underline{4}.882$ 20. $2\underline{0}8.96$ 21. $0.004\underline{5}4$

Estimate using rounding (round to the nearest whole number).

22. $352.67 + 40.098 \approx$ 23. $82.47 - 9.728 \approx$

24. $24.777 + 35.22 \approx$ 25. $19.543 - 10.453 \approx$

Lesson 4 Homework 3

Round to the nearest whole number.

1. 22.72   2. 0.499   3. 7.7111

Round to the nearest tenth.

4. 0.228   5. 6.652   6. 0.071

Round to the nearest hundredth.

7. 2.387   8. 0.0127   9. 16.707

Round to the nearest thousandth.

10. 34.0234   11. 0.66667   12. 0.00119

Round to the nearest ten-thousandth.

13. 1.23456   14. 0.0000333   15. 22.098765

Round to the place of the underlined digit.

16. $0.8\underline{7}43$ 17. $3\underline{9}9.7$ 18. $24.\underline{8}09$

19. $0.0\underline{0}85$ 20. $\underline{5}.607$ 21. $39.67\underline{2}6$

Estimate using rounding (round to the nearest whole number).

22. $73.68 + 26.92 \approx$ 23. $25.11 - 3.7 \approx$

24. $3.624 + 29.483 \approx$ 25. $996.7 - 324.22 \approx$

Lesson 5 Homework 1

Add or subtract.

1. $\begin{array}{r} 0.275 \\ +\ \underline{0.783} \end{array}$

2. $\begin{array}{r} 6.209 \\ -\ \underline{0.48} \end{array}$

3. $\begin{array}{r} 24.03 \\ +\ \underline{8.28} \end{array}$

4. $\begin{array}{r} 50.05 \\ -\ \underline{8.43} \end{array}$

5. $\begin{array}{r} 0.86 \\ +\ \underline{9.208} \end{array}$

6. $\begin{array}{r} 0.0034 \\ -\ \underline{0.00127} \end{array}$

7. $\begin{array}{r} 39.29 \\ +\ \underline{9.45} \end{array}$

8. $\begin{array}{r} 12.065 \\ -\ \underline{8.28} \end{array}$

9. $\begin{array}{r} 283.37 \\ +\ \underline{99.286} \end{array}$

10. 23.45 + 7.926 =

11. 68.93 - 19.68 =

12. 0.234 + 12.32 =

13. 50 - 9.257 =

14. 17.034 + 6.28 =

15. 0.896 - 0.698 =

16. 24.5 + 3.78 =

17. 32.08 - 16.3 =

18. 2.3 + 0.44 + 12.8 =

19. 1.2 - 0.837 =

20. 43.28 + 8.037 + 125.9 =

21. 207.892 + 38.07 + 600.091 =

22. 1000 - 208.32 - 36.77 =

23. 48.12 - 9.24 - 12.543 =

24. 65.27 - 23.45 + 56.048 =

Lesson 5 Homework 2

Add or subtract.

1. 0.37 + 0.78
2. 0.078 - 0.049
3. 6.809 + 0.79
4. 23.06 - 16.68
5. 3.247 + 9.76
6. 0.306 - 0.087
7. 84.86 + 27.9
8. 4.092 - 1.496
9. 21.23 + 157.048

10. 0.098 + 23.64 =

11. 20.06 - 8.98 =

12. 6.9 + 29.208 =

13. 102.07 - 63.385 =

14. 24.24 + 76.76 =

15. 3.09 - 0.983 =

16. 0.023 + 20.98 =

17. 12 - 11.543 =

18. 6.49 + 0.744 =

19. 0.808 - 0.79 =

20. 59.73 + 8.9 + 12.84 =

21. 6.087 + 7.49 + 12.965 =

22. 69.9 - 27.83 - 9.67 =

23. 90.84 - 25.93 - 9.045 =

24. 8.37 - 3.78 + 4.296 =

Lesson 6 Homework 1

Word problems.

1. Jim spent $22.50 buying jeans and $3.75 less than that buying a T-shirt. How much did he spend altogether?

2. Jennifer had a $20 bill. She bought a chocolate bar for $3.35 and an apple pie for $7.90. How much does she have left?

3. Ann is 1.53 meters tall. Her sister is 0.27 meters shorter, and her brother is 0.18 meters taller. How tall are the sister and the brother?

4. To repair his house, Mr. Black needs a wooden stud 0.78 meters long. In a store he could find studs only 0.65 or 0.87 meters long. Which size is closer to what he needs?

5. A rectangular lot is 25.8 feet long and 12.35 feet wide. Find the perimeter of the lot.

6. The perimeter of a triangle is 62.3 meters. One of the sides is 22.8 meters; the second one is 17.9 meters. Find the third side.

7. Larry decided to buy a compact disc. In a store, he found out that if one buys one disc, the second is sold at $4.50 off. How much did he pay for two discs if the price for the first disc is $11.25?

8. A family of four, two parents and two kids, went to see a movie. An adult's ticket is $6.50; a kid's ticket is $1.75 cheaper. How much did the family pay to see a movie?

9. To rent a boat you should pay $15 for the first hour, and then $2.30 less for each additional hour. How much should you pay if you want to rent a boat for 3 hours?

10. Mary called to her relatives in London and talked 5.3 minutes with her mother, 2.9 minutes with her cousin, and 8.4 minutes with her grandmother. How long was her international call?

Chapter 7 Test

Write as a decimal

1. $\frac{7}{10}$ 2. $\frac{23}{100}$ 3. $3\frac{37}{1000}$ 4. $7\frac{46708}{100000}$

Write in words and in expanded form

5. 0.045 6. 2.34875 7. 12.007532

Make a number line and show each decimal on the number line

8. 0.6 9. 1.3 10. 1.85 11. 2.15

Write >, <, or = for each □

12. 0.409 □ 0.41 13. 6.311 □ 6.2999 14. 0.520 □ 0.52

Round to the place of the underlined digit.

15. $0.\underline{8}5$ 16. $1.2\underline{3}49$ 17. $7\underline{2}.8421$ 18. $0.00\underline{5}218$

Add or subtract

19. 0.32 + 2.7 =
20. 3.5 – 0.782 =
21. 25.03 + 7.378 =
22. 62.04 – 9.396 =
23. 7.7 + 24.28 + 0.937 =
24. 70.2 – 14.22 – 27.673 =

Word problems

25. Kim went to a bookstore and bought three textbooks: Algebra, Geometry, and Trigonometry. The algebra textbook costs \$32.57, the geometry textbook \$47.99, and the trigonometry textbook \$28.25. How much money did Kim pay for her good mathematical education?

26. Gwen's weight is 123.5 pounds. Her daughter's weight is 45.75 pounds less. How much do they weigh together?

Chapter 8

Lesson 1 Homework 1

Multiply.

1. 4.8 x 0.7

2. 0.24 x 0.3

3. 31.7 x 6

4. 1.7 x 4.6

5. 0.45 x 2.06

6. 2.04 x 4.9

7. 50.8 x 1.2

8. 3.44 x 3.7

9. 0.045 x 12

10. 0.08 x 10 =

11. 6.23 x 100 =

12. 13.5 x 100 =

13. 0.867 x 1000 =

14. 6.04 x 0.35 =

15. 0.078 x 2.6 =

16. 21.5 x 3.7 =

17. 0.08 x 0.06 =

18. 16.02 x 45 =

19. 67.035 x 80 =

Estimate, then find the product.

20. 5.8 x 8.2 ≈

21. 28.92 x 3.05 ≈

22. 0.9 x 7.8 ≈

23. 82.3 x 40.6 ≈

24. 24.8 x 9.9 ≈

25. 1.2 x 1.32 ≈

Lesson 1 Homework 2

Multiply.

1. 0.07 x 0.08

2. 3.4 x 8.9

3. 3.56 x 4.5

4. 7.2 x 0.005

5. 4.28 x 4.6

6. 18.22 x 0.07

7. 8.234 x 0.7

8. 0.0024 x 108

9. 68 x 0.04

10. 32.4 x 10 =

11. 0.048 x 100 =

12. 0.075 x 10 =

13. 65.78 x 1000 =

14. 0.012 x 8.6 =

15. 0.6 x 0.0015 =

16. 1.58 x 0.004 =

17. 82.6 x 0.0025 =

18. 0.007 x 0.009 =

19. 200 x 0.0048 =

Estimate, then find the product.

20. 4.8 x 9.3 =

21. 24.9 x 0.09

22. 3.08 x 5.09 =

23. 160.5 x 0.87

24. 98.3 x 4.8 =

25. 0.05 x 602.4

Lesson 2 Homework 1

Divide. Check the answer.

1. $2\overline{)8.46}$
2. $3\overline{)4.17}$
3. $4\overline{)12.12}$
4. $5\overline{)21.5}$
5. $6\overline{)0.354}$
6. $7\overline{)4.564}$
7. $3\overline{)14.25}$
8. $9\overline{)1.863}$
9. $5\overline{)0.0665}$
10. $8\overline{)25.672}$
11. $4\overline{)64.32}$
12. $7\overline{)0.231}$
13. $11\overline{)2.673}$
14. $9\overline{)0.981}$
15. $12\overline{)70.8}$
16. $18\overline{)42.12}$
17. $6\overline{)0.0126}$
18. $21\overline{)1.764}$
19. $8\overline{)0.3472}$
20. $7\overline{)42.343}$
21. $4\overline{)355.6}$
22. $25\overline{)0.625}$
23. $24\overline{)5.76}$
24. $3\overline{)89.34}$
25. $14\overline{)2.198}$
26. $31\overline{)21.08}$
27. $9\overline{)1.116}$

Lesson 2 Homework 2

Divide. Check your answer.

1. 3) 1.338	2. 5) 0.275	3. 4) 67.16
4. 2) 79.92	5. 7) 2.478	6. 6) 9.54
7. 8) 0.0784	8. 10) 7.890	9. 9) 0.657
10. 12) 1.356	11. 14) 5.614	12. 23) 0.828
13. 17) 42.16	14. 22) 4.796	15. 7) 609.7
16. 11) 36.63	17. 41) 0.574	18. 8) 0.0752
19. 35) 19.25	20. 12) 4.092	21. 9) 19.35
22. 16) 0.384	23. 19) 1.995	24. 3) 0.1284
25. 15) 33.15	26. 20) 2.480	27. 5) 0.0345

Lesson 3 Homework 1

Write each repeating decimal using bar notation.

1. 0.777... =
2. 0.232323... =
3. 2.7838383... =
4. 6.666... =
5. 56.247247... =
6. 0.030303... =

Divide.

7. $3\overline{)\,0.7}$
8. $6\overline{)\,3.5}$
9. $9\overline{)\,4.7}$
10. $11\overline{)\,0.5}$
11. $12\overline{)\,70}$
12. $6\overline{)\,77.2}$
13. $3\overline{)\,25.25}$
14. $7\overline{)\,0.1}$
15. $9\overline{)\,2.2}$

Change each fraction to a decimal.

16. $\frac{1}{6} =$
17. $\frac{20}{11} =$
18. $\frac{8}{9} =$
19. $\frac{13}{8} =$
20. $12\frac{3}{4} =$
21. $3\frac{5}{6} =$
22. $\frac{6}{7} =$
23. $5\frac{11}{12} =$
24. $\frac{45}{16} =$

Lesson 3 Homework 2

Write each repeating decimal using bar notation.

1. 0.0555...
2. 7.37373...
3. 22.3453453...
4. 0.07666...
5. 787.8787...
6. 0.0340340...

Divide.

7. $6\overline{)\,0.5}$
8. $3\overline{)\,20.3}$
9. $7\overline{)\,2}$
10. $9\overline{)\,15.7}$
11. $12\overline{)\,2.5}$
12. $11\overline{)\,0.3}$
13. $3\overline{)\,0.02}$
14. $15\overline{)\,50}$
15. $21\overline{)\,45.5}$

Change each fraction to a decimal.

16. $\frac{2}{3} =$
17. $\frac{7}{6} =$
18. $\frac{52}{9} =$
19. $\frac{3}{8} =$
20. $5\frac{2}{7} =$
21. $20\frac{8}{11} =$
22. $\frac{81}{16} =$
23. $\frac{14}{15} =$
24. $\frac{35}{14} =$

Lesson 4

Homework 1

Divide. Round to the nearest whole number.

1. $5\overline{)8}$
2. $6\overline{)200}$
3. $9\overline{)176}$
4. $12\overline{)13}$
5. $20\overline{)39}$
6. $15\overline{)763}$

Divide. Round to the nearest tenth.

7. $7\overline{)5.4}$
8. $3\overline{)25}$
9. $4\overline{)0.7}$
10. $8\overline{)25.789}$
11. $11\overline{)0.3}$
12. $9\overline{)853}$

Divide. Round to the nearest hundredth.

13. $3\overline{)1.34}$
14. $4\overline{)0.11}$
15. $6\overline{)37.1}$
16. $12\overline{)92.5}$
17. $7\overline{)2.7}$
18. $16\overline{)0.2}$

Divide. Round to the nearest thousandth.

19. $8\overline{)0.62}$
20. $9\overline{)291}$
21. $15\overline{)9.5}$
22. $7\overline{)25.3}$
23. $11\overline{)47.47}$
24. $6\overline{)0.045}$

Divide. Round to the nearest ten-thousandth.

25. $3\overline{)0.67}$
26. $13\overline{)0.072}$
27. $9\overline{)4.44}$

Lesson 4 Homework 2

Divide. Round to the nearest whole number.

1. $9\,\overline{)\,12}$
2. $3\,\overline{)\,26}$
3. $4\,\overline{)\,0.93}$
4. $24\,\overline{)\,79}$
5. $11\,\overline{)\,356}$
6. $6\,\overline{)\,17}$

Divide. Round to the nearest tenth.

7. $3\,\overline{)\,4}$
8. $7\,\overline{)\,29.345}$
9. $14\,\overline{)\,0.55}$
10. $20\,\overline{)\,37.1}$
11. $9\,\overline{)\,0.17}$
12. $24\,\overline{)\,371.2}$

Divide. Round to the nearest hundredth.

13. $6\,\overline{)\,7}$
14. $7\,\overline{)\,0.8}$
15. $23\,\overline{)\,48.5}$
16. $30\,\overline{)\,0.723}$
17. $9\,\overline{)\,21.9}$
18. $8\,\overline{)\,0.3456}$

Divide. Round to the nearest thousandth.

19. $7\,\overline{)\,10}$
20. $3\,\overline{)\,0.74}$
21. $9\,\overline{)\,2.723}$
22. $15\,\overline{)\,0.47}$
23. $12\,\overline{)\,8.321}$
24. $30\,\overline{)\,25}$

Divide. Round to the nearest ten-thousandth.

25. $6\,\overline{)\,0.874}$
26. $7\,\overline{)\,3.3}$
27. $8\,\overline{)\,21.1}$
28. $31\,\overline{)\,0.5}$
29. $16\,\overline{)\,89}$
30. $11\,\overline{)\,0.32}$

Lesson 5

Homework 1

Divide. Check the answer.

1. $0.3 \overline{)1.2}$
2. $0.5 \overline{)0.04}$
3. $0.12 \overline{)3.6}$
4. $0.7 \overline{)0.049}$
5. $0.06 \overline{)8.4}$
6. $0.15 \overline{)0.087}$
7. $1.7 \overline{)3.519}$
8. $2.5 \overline{)0.625}$
9. $0.006 \overline{)9}$
10. $0.003 \overline{)0.00003}$
11. $0.012 \overline{)0.0000456}$
12. $0.0032 \overline{)17.92}$
13. $0.0014 \overline{)4.76}$

Find each quotient to the nearest tenth.

14. $3.5 \overline{)0.868}$
15. $0.24 \overline{)0.1872}$

Find each quotient to the nearest whole number.

16. $0.0025 \overline{)0.762}$
17. $0.4 \overline{)0.392}$
18. $0.045 \overline{)9.27}$
19. $0.03 \overline{)0.009}$

Estimate each quotient.

20. $6.3 \overline{)20.5}$
21. $0.24 \overline{)1.6}$
22. $0.0049 \overline{)0.0512}$
23. $1.8 \overline{)0.208}$

Lesson 5 Homework 2

Divide. Check the answer.

1. $0.4\overline{)21.6}$
2. $0.07\overline{)16.1}$
3. $0.31\overline{)0.465}$
4. $2.2\overline{)0.726}$
5. $0.09\overline{)3.06}$
6. $1.5\overline{)1.095}$
7. $0.032\overline{)76.8}$
8. $1.03\overline{)0.2266}$
9. $0.92\overline{)2.392}$
10. $0.008\overline{)0.0000016}$
11. $0.0027\overline{)8.1}$
12. $0.0103\overline{)0.003502}$
13. $0.056\overline{)2.744}$

Find each quotient to the nearest tenth.

14. $0.8\overline{)33.5}$
15. $0.038\overline{)6.47}$
16. $0.025\overline{)0.357}$
17. $0.\ 0057\overline{)0.223}$

Find each quotient to the nearest whole number.

18. $0.045\overline{)3.008}$
19. $0.83\overline{)62.4}$
20. $0.006\overline{)37.21}$
21. $0.0087\overline{)0.054}$

Estimate each quotient.

22. $5.5\overline{)23.92}$
23. $0.048\overline{)0.987}$
24. $0.093\overline{)12.4}$
25. $24.7\overline{)10.08}$

Lesson 5 Homework 3

Divide and check

1) $2.8 \div 0.8 =$

2) $1.8 \div 0.24 =$

3) $7.35 \div 0.35 =$

4) $6.02 \div 0.7 =$

5) $1.35 \div 2.5 =$

6) $72.6 \div 0.33 =$

7) $0.1218 \div 0.6 =$

8) $42.9 \div 6.6 =$

9) $0.2222 \div 0.44 =$

10) $0.085 \div 6.8 =$

11) $0.468 \div 0.06 =$

12) $50.82 \div 7.7 =$

13) $2.256 \div 0.24 =$

14) $0.22644 \div 0.333 =$

15) $2{,}041.2 \div 4.05 =$

16) $1.036 \div 0.35 =$

Lesson 6 Homework 1

Word problems.

1. Sharla works as a babysitter 6 days a week, 3.5 hours a day. How much does she earn in an hour if she got $130.25 during the week including a $20 bonus?

2. For a new school year Larry bought 5 notebooks, 3 pens, and 7 pencils. How much did he pay if each notebook costs $1.25, each pen $2.63, and each pencil $0.24?

3. Jacob decided to visit his friend Jim. He drove 2.7 hours with average speed 57.3 miles per hour. How far apart do the friends live?

4. Mrs. Smith bought two pairs of jeans and six T-shirts for her twins. How much did she spend if the jeans cost $ 25.80, and each T-shirt is twice as cheap?

5. Gwen jogs 3 miles every morning. What is her average speed if she spends 25 minutes for jogging? (Speed should be in miles per minute).

6. Jonathan sent 8 greeting cards to his friends in the USA and 5 cards to his friends abroad. Each card in the USA cost $0.75, and each card abroad cost $1.25. What was more expensive: to send 5 cards abroad or 8 cards in the USA?

7. 0.25 of every hour of TV airtime is spent on commercials. If a child spent 27 hours a week watching TV, how much time did he spend watching commercials?

8. Max started reading a book at 9:00 a.m. and finished at 3:00 p.m. How many pages did he read per minute if the book had 450 pages?

(hint: 1 hour = 60 minutes).

9. What is the average of these numbers: 5, 7, 8, 11?

10. To sew one dress, Mrs. Larsen needs 2.25 meters of fabric. How many dresses can she make from 30 meters?

Chapter 8 Test

Multiply

1. $\begin{array}{r} 3.4 \\ \times\ 0.8 \\ \hline \end{array}$
2. $\begin{array}{r} 0.47 \\ \times\ 3.2 \\ \hline \end{array}$
3. $\begin{array}{r} 24.3 \\ \times\ 0.87 \\ \hline \end{array}$
4. $\begin{array}{r} 0.09 \\ \times\ 0.06 \\ \hline \end{array}$

5. 29.53 x 10 =
6. 3.058 x 100 =
7. 0.12 x 7.03 =
8. 0.25 x 6.24 =

Divide

9. $2\ \overline{)\ 0.126}$
10. $7\ \overline{)\ 2.345}$
11. $12\ \overline{)\ 62.76}$
12. $0.3\ \overline{)\ 0.252}$
13. $0.12\ \overline{)\ 4.92}$
14. $0.008\ \overline{)\ 0.2}$

Change each fraction to a decimal. Use bar notation if necessary.

15. $\frac{1}{3} =$
16. $\frac{5}{6} =$
17. $\frac{35}{8} =$
18. $7\frac{11}{12} =$

Find each quotient to the nearest tenth

19. $0.7\ \overline{)\ 4.6783}$
20. $2.2\ \overline{)\ 0.84}$
21. $9\ \overline{)\ 41}$

Word Problems

22. Peter bought 3 pencils and paid \$0.75 for them. Adam bought a package with 8 pencils and paid \$1.8 for it. How much cheaper is a pencil in the package than a single pencil?

23. Mr. Goodman went to a flower shop to buy flowers for his wife. He bought 6 roses, 4 carnations, and 10 tulips. How much did he pay for the flowers if each rose costs \$2.25, each carnation costs \$1.75, and each tulip \$0.75?

Chapter 9

Lesson 1 Homework 1

Evaluate.

1. $1-\frac{1}{2}\times\frac{4}{5}=$

2. $2\frac{1}{3}+3\frac{1}{2}\div5\frac{1}{4}=$

3. $\frac{5}{6}\times\frac{9}{10}-\frac{1}{2}\times\frac{3}{4}=$

4. $\frac{1}{2}\times\left(\frac{1}{3}+\frac{1}{4}\right)=$

5. $\frac{2}{5}+\frac{3}{4}\times\frac{2}{5}=$

6. $2\times\left(3\frac{1}{2}-1\frac{1}{4}\right)=$

7. $\frac{1}{2}\times\frac{2}{3}+\frac{1}{4}\times\frac{2}{3}+\frac{1}{3}\times\frac{1}{2}=$

8. $7\frac{4}{5}-\frac{4}{5}\times\frac{1}{2}+3\frac{1}{2}=$

9. $3\frac{1}{2}+3\times\left(4\frac{1}{3}-3\frac{2}{3}\right)=$

10. $2\frac{1}{2}\div\left(\frac{1}{2}+\frac{3}{4}\right)+5\frac{1}{7}=$

Lesson 1 Homework 2

Evaluate.

1. $\frac{3}{5}+\frac{4}{7}\times\frac{7}{10}=$

2. $10-2\frac{1}{4}\times3\frac{1}{3}=$

3. $\frac{8}{9}\div1\frac{1}{7}-\frac{7}{10}\div\frac{14}{15}=$

4. $3\times\left(\frac{2}{3}+\frac{2}{5}\times1\frac{2}{3}\right)=$

5. $1\frac{7}{8}\times5\frac{1}{3}-6\frac{2}{3}\div3\frac{1}{3}=$

6. $9\frac{1}{2}-6\times\left(\frac{1}{3}-\frac{1}{4}\right)=$

7. $1\frac{5}{7}\times4\frac{2}{3}+3\frac{3}{4}\times5\frac{1}{3}=$

8. $\frac{2}{3}\times3+\frac{3}{5}\times5-\frac{4}{7}\times7=$

9. $\frac{2}{5}\times6\frac{1}{4}\times1\frac{3}{5}+\frac{3}{7}\times2\frac{1}{3}=$

10. $100-2\frac{1}{2}\times3\frac{1}{5}\times4\frac{1}{4}=$

Lesson 2 Homework 1

Evaluate.

1. $1.2 + 0.6 \times 0.5 =$

2. $7.3 - 3.3 \div 1.1 =$

3. $(6.5 + 3.5) \times 0.07 =$

4. $2.5 \times 0.4 - 3.6 \times 0.2 =$

5. $0.1 \times (5.42 - 0.32) =$

6. $12.4 - 4 \times (0.7 + 0.4) =$

7. $\dfrac{7.28 + 2.72}{0.4 \times 5} =$

8. $\dfrac{2.8 + 4.4 \times 0.5}{7.2 - 6.6 \div 3} =$

9. $\dfrac{0.2 \times 1.5 + 1.2 \div 0.4}{0.7 + 0.4} =$

10. $\dfrac{3.8 \times (1.26 - 1.06)}{2.4 \times 5 - 2} =$

11. $0.0032 \div 0.00008 \times 0.025 =$

12. $0.014 \div 0.0007 \div 0.04 =$

Lesson 2 Homework 2

Evaluate.

1. $5.6 - 1.2 \times 0.5 =$

2. $4.4 \div 0.22 + 3.4 \times 1.5 =$

3. $12.25 + 0.3 \times 2.5 =$

4. $4.8 \div 12 - 0.08 =$

5. $0.18 + 2 \times (0.03 + 0.26) =$

6. $0.2 \times 0.7 - 0.2 \times 0.07 =$

7. $\dfrac{12.65 + 2.35}{0.3 + 0.1} =$

8. $\dfrac{3.92 + 0.4 \times 0.2}{0.2 \times 0.2} =$

9. $\dfrac{18 \div 0.45 - 14 \div 0.35}{6 \div 0.03} =$

10. $\dfrac{2.6 \times (1.2 - 0.2 \times 3.5)}{6 \times 0.2 + 0.1} =$

11. $0.00036 \div 0.012 \div 0.005 =$

12. $0.0035 \div (0.002 + 0.003) + 0.3 =$

Lesson 3 Homework 1

Evaluate.

1. $5.5+\frac{1}{4}\times\frac{4}{5}=$

2. $\frac{7}{8}-0.2\times 0.3=$

3. $15-0.8\div\frac{2}{3}=$

4. $6.4\times\frac{5}{8}+14\times 0.2=$

5. $0.3\div 0.03-\frac{3}{7}\div 1\frac{1}{14}=$

6. $18-3.6\times\left(\frac{1}{2}-\frac{1}{4}\right)=$

7. $0.2\times\left(\frac{7}{8}-\frac{3}{4}\right)+\frac{2}{5}=$

8. $0.4\times 0.5\times 0.6+\frac{2}{3}\times\frac{3}{4}\times\frac{4}{5}=$

9. $\dfrac{3.2-2\times 1.45}{\frac{1}{4}+\frac{1}{5}}=$

10. $\dfrac{\frac{2}{3}\div 1\frac{2}{3}+0.8\div 4}{1-0.6}=$

Lesson 3 Homework 2

Evaluate.

1. $\frac{2}{3}\times 0.6+\frac{3}{4}\times 0.8=$

2. $3\div 0.75-\frac{5}{6}\times 0.9=$

3. $25-0.08\div 0.005=$

4. $5\frac{1}{2}\times 0.4+7.2\div 2\frac{2}{5}=$

5. $3.6\times\left(1\frac{1}{2}+4\frac{1}{3}\right)=$

6. $21\frac{1}{10}-3.4\times\left(\frac{3}{8}-\frac{1}{4}\right)=$

7. $0.8\times 0.9\div 7\frac{1}{5}=$

8. $2.4\times\left(\frac{3}{4}+\frac{5}{6}\right)-2\frac{1}{7}=$

9. $\dfrac{6.6-\frac{3}{4}\times\frac{4}{5}}{3.2-\frac{2}{3}\times 0.3}=$

10. $\dfrac{5\frac{1}{3}+9\frac{3}{5}\div 3.6}{0.4\times 0.002}=$

Lesson 4 Homework 1

Word problems.

1. The sum of three consecutive whole numbers is 36. Find the numbers.

2. Max finished reading a very interesting book and ran with the speed of 5 miles per hour to his friend Ezra to give it to him. Ezra couldn't wait and ran toward Max with a speed of 6 miles per hour. How far apart do the friends live, if in a half an hour Ezra already started reading the book?

3. Mrs. Brown bought 40 candies for her 3 kids. On the way home she has eaten $\frac{1}{4}$ of the candies, but then she divided what was left equally between her kids. How many candies did each kid get?

4. Cindy, Mary, and Betty spent \$360 together. Cindy spent only $\frac{1}{8}$ of this amount, but Betty spent $\frac{2}{3}$. How much money did Mary spend?

5. A subscription to a children's magazine costs \$15 for 12 issues. How much will you save by subscribing if a single magazine costs \$1.45?

6. Rita practices the piano $\frac{3}{4}$ of an hour each day 5 times a week. Her teacher wants her to practice 1.5 hour every day and 7 times a week. How many more hours per week should Rita practice to please her teacher?

7. To decorate for a party, the Johnsons bought 60 balloons. They noticed that they bought 5 more red balloons than blue ones, and 5 more blue ones than white ones. How many red, blue and white balloons do the Johnsons have?

8. The fifth grade class had a math test. $\frac{3}{4}$ of the students were successful, but 7 students failed the test. How many students are in the class?

9. If I reduce a fraction, I will get $\frac{2}{3}$. The sum of the numerator and the denominator is 20. What is the fraction?

10. 50 students participated in a parade. There were 8 more boys than girls. How many boys and how many girls participated in the parade?

Chapter 9 Test

Evaluate.

1. $25 - 5 \times 3\frac{1}{5} =$
2. $6\frac{3}{4} - \frac{3}{4} \times \frac{2}{3} =$
3. $\frac{4}{5} \div \frac{8}{15} - \frac{7}{8} \times \frac{4}{7} =$
4. $12 \times \left(\frac{1}{3} + \frac{1}{4}\right) + 6 \div \frac{2}{7} =$
5. $3.8 + 1.5 \times 6.4 =$
6. $6.4 \div 0.08 - 0.5 \div 0.025 =$
7. $0.15 - 0.2 \times (0.7 + 0.05) =$
8. $\frac{2.2 + 0.5 \times 3.6}{0.4 \times 0.5} =$
9. $0.3 \div \frac{3}{4} - 0.2 \times \frac{5}{6} =$

Word problem

11. Jonathan got $1,000 for his summer job. He deposited $\frac{3}{4}$ of this amount to a bank, and decided to spend what was left. He bought 2 pairs of jeans, $54.99 each, and 3 T-shirts, $24.45 each. How much does he have left?

Chapter 10

Lesson 1 Homework 1

Give the absolute value.

1. | -8 | 2. | 15 | 3. | -52 |

4. | -41 | 5. | 0 | 6. | 125 |

Give the opposites of the numbers.

7. 23 8. -14 9. -100

10. | -44 | 11. 73 12. | 10 |

Give the integer for each point on the number line.

```
______.____.____.____.____.____.____|____.____.____.____.____.____.____.______
      A         B    C         -1   0    1         D         E
```

13. A 14. B 15. C 16. D 17. E

Give the integer for the point on the number line.

18. A point 13 units to the left of 0

19. A point 20 units to the right of -8

20. A point 8 units to the left of | -3 |

Lesson 2 Homework 1

Write <, >, or = for each □.

1. -7 □ 7
2. 14 □ -16
3. -28 □ -17
4. | -23 | □ 23
5. -40 □ -54
6. -50 □ | -75 |

Order the numbers from least to greatest.

7. 9, -3, 0, | - 16 |, -12
8. -7, | -22 |, 45, -38, 7

Order the numbers from greatest to least.

9. -15, 12, | -36 |, -8, 2
10. 40, -30, 70, -20, | -50 |

Use the number line to plot points representing the integers in the parentheses. Then write them in order from least to greatest.

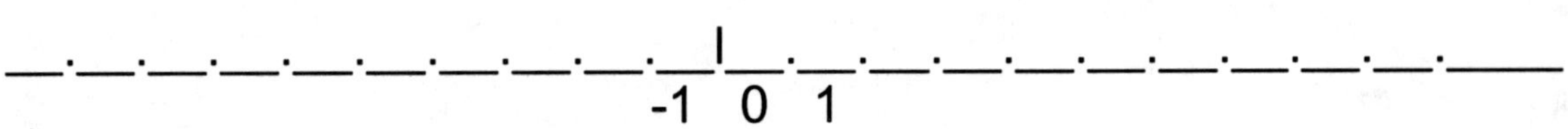

11. A (2)
12. B (-5)
13. C (7)
14. D (-2)

Give the integer for the point on the number line.

15. A point 6 units to the left of 0
16. A point 12 units to the right of -7
17. A point 10 units to the left of 5

Lesson 3 Homework 1

Find each sum.

1. -7 + (-12) =
2. 13 + (-18) =
3. -25 + 17 =
4. 28 + 33 =
5. 0 + (-59) =
6. -123 + (- 58) =
7. 48 + (-52) =
8. -100 + 77 =
9. 45 + (-45) =
10. 56 + 64 =
11. -17 + (-38) =
12. 29 + (-85) =
13. -41 + 128 =
14. -83 + (-49) =
15. 120 + (-97) =
16. -44 + 152 =
17. 69 + 73 =
18. -36 + (-76) =
19. -82 + 0 =
20. 65 + (-147) =
21. -111 + (-19) =
22. -72 + 200 =
23. -125 + 48 =
24. -99 + (-99) =
25. 160 + (-78) =
26. 36 + (-36) =
27. -39 + (- 139) =
28. -53 + 121 =
29. 129 + 234 =
30. -147 + 88 =

Lesson 4 Homework 1

Find each difference.

1. 8 - 12 =
2. -3 - 5 =
3. 15 - 3 =
4. -18 - (-3) =
5. 24 - 32 =
6. 9 - (- 11) =
7. -16 - 15 =
8. -6 - (- 14) =
9. 45 - 18 =
10. 0 - 35 =
11. 16 - 33 =
12. -28 - 34 =
13. 40 - (- 23) =
14. -46 - 0 =
15. -22 - (-38) =
16. -18 - (-18) =
17. -17 - 87 =
18. 56 - 57 =
19. 24 - 49 =
20. 15 - (- 16) =
21. -40 - (- 68) =
22. -92 - 29 =
23. 106 - 112 =
24. 210 - 87 =
25. 35 - (-77) =
26. -103 - (-63) =
27. -49 - 94 =
28. 36 - 81 =
29. 324 - 187 =
30. 62 - (-38) =

Lesson 4 Homework 2

Add or subtract.

1. 8 + (-24) =
2. -24 - 38 =
3. -25 - 42 =
4. -28 + 62 =
5. 28 - (-38) =
6. 45 + (-28) =
7. -19 + (-42) =
8. 12 - 58 =
9. -45 + 45 =
10. -32 - (-48) =
11. -31 + 62 =
12. -127 + 49 =
13. 14 - (-48) =
14. 65 - 142 =
15. -28 + (-56) =
16. -32 - 62 =
17. -9 - (-91) =
18. 51 + (-36) =
19. 57 - 38 =
20. 100 - (-77) =
21. -22 + 80 =
22. -46 + (-54) =
23. -62 - 28 =
24. 17 - 104 =
25. 41 + (-76) =
26. -25 + 75 =
27. 19 - (-46) =
28. -29 + (-65) =
29. -96 + 38 =
30. -61 - (- 88) =

Lesson 4 Homework 3

Add or subtract.

1. 14 - 28 =
2. 35 + (-43) =
3. -56 + (- 35) =
4. -26 - 48 =
5. 25 - (-43) =
6. 72 - 38 =
7. 29 + 58 =
8. -17 - (-53) =
9. -31 - 11 =
10. -42 + (-48) =
11. 36 - 65 =
12. 82 + (-37) =
13. 24 + (-52) =
14. -48 - 35 =
15. 14 - 92 =
16. -72 + 19 =
17. (-64) + (-11) =
18. 28 - (- 57) =
19. -47 - (- 53) =
20. 94 - 48 =
21. 132 + (- 68) =
22. -22 + 78 =
23. -73 - 87 =
24. -112 - (-68) =
25. 88 - (-48) =
26. -55 + (-65) =
27. 143 + 257 =
28. 33 - 99 =
29. -69 - (-73) =
30. -148 + 72 =

Lesson 5

Homework 1

Find the product.

1. -8 x 12 =
2. 13 x (-6) =
3. -9 x (-20) =
4. 15 x 15 =
5. 14 x (-25) =
6. -4 x (-36) =
7. -11 x (-11) =
8. -65 x 3 =
9. 28 x 16 =
10. 21 x (-14) =
11. -22 x (-9) =
12. -8 x 19 =
13. 28 x (-7) =
14. -18 x (-19) =
15. -13 x 52 =
16. 30 x (-40) =
17. 24 x (-5) =
18. 42 x 25 =
19. -38 x (-35) =
20. -54 x 16 =
21. -3 x (- 4) x (-5) =
22. -4 x 5 x (-7) =
23. 8 x (-9) x 10 =
24. 6 x 7 x (-9) =
25. -2 x 8 x 5 =
26. -6 x (-3) x 14 =
27. -5 x (-12) x (- 7) =
28. 3 x (-11) x 8 =
29. -1 x (-2) x (-3) x (-4) =
30. (-234) x (-45) =

Lesson 6

Homework 1

Find the quotient.

1. 28 ÷ (-7) =
2. -56 ÷ (-8) =
3. 27 ÷ 9 =
4. -42 ÷ 6 =
5. -35 ÷ (-5) =
6. 400 ÷ (-4) =
7. 0 ÷ (-13) =
8. -66 ÷ (-22) =
9. -60 ÷ 12 =
10. 640 ÷ 80 =
11. 75 ÷ (-5) =
12. -51 ÷ 17 =
13. -81 ÷ (-9) =
14. 120 ÷ (-24) =
15. 68 ÷ (-4) =
16. -366 ÷ (-6) =
17. 96 ÷ 12 =
18. -72 ÷ 8 =
19. -121 ÷ 11 =
20. 169 ÷ (-13) =
21. 12 ÷ (-2) ÷ (-3) =
22. -80 ÷ (-16) =
23. -90 ÷ 10 ÷ (-3) =
24. -480 ÷ 60 =
25. -24 ÷ (-6) ÷ (-2) =
26. 300 ÷ (-75) =
27. 100 ÷ (-4) ÷ (-5) =
28. -225 ÷ (-15) =
29. 360 ÷ (-6) ÷ 4 =
30. -180 ÷ 36 =

Lesson 7 Homework 1

Evaluate.

1. $12 - 3 \times (-2) =$
2. $-4 \times 3 + 7 =$
3. $-18 \div (-6) - (-4) =$
4. $7 \times (-12 - 8) =$
5. $-4 \times (6 - 10) =$
6. $2 \times (-3) + (-5) \times 4 =$
7. $-16 \div 8 - 14 =$
8. $-24 \div (-6) \times (-9) =$
9. $-11 - 12 \div (-4) =$
10. $15 - 3 \times (7 - 11) =$
11. $-6 - (-7) \times (-8) =$
12. $-12 \div 6 + 18 \div (-3) =$
13. $20 \div (-4) \times (7 - 12) =$
14. $-2 \times (-5) - 18 =$
15. $32 \div (-4) \div (-2) - (-14) =$
16. $-5 \times 7 + 24 \div (-6) =$

Lesson 7 Homework 2

Evaluate.

1. $4 \times [13 - (-7)] =$
2. $25 \div (-5) + (-20) =$
3. $-6 \times (-8) - 52 =$
4. $-25 - 3 \times (-12) =$
5. $3 + 2 \times (-16 + 4) =$
6. $75 \div [3 \times (4 - 9)] =$
7. $-16 - 12 \div (-4) =$
8. $-2 + (-2) \times (-2 - 2) =$
9. $4 \times [4 - 4 \times (5 - 9)] =$
10. $2 \times (-3) - 3 \times (-2) =$
11. $-14 + (-3) \times (-7) =$
12. $-81 \div (-9) \div (-3) - 6 =$
13. $12 + (-12) \div 4 - 8 \div 2 =$
14. $-23 - 3 \times (-2) - (-9) =$
15. $42 \div (-6) - 48 \div (-8) =$
16. $30 + (-3) \times (-15 - 6) =$

Lesson 8 Homework 1

Solve word problems.

1. Mr. Williams opened a checking account and deposited $500. Then he withdrew $340 and after that he withdrew $180. What is his balance after these two transactions?

2. Michael's math teacher gives 2 extra points for solving each extra credit problem, and takes away 5 points for each incorrect answer. Michael solved 5 problems incorrectly, but then he solved 2 extra credit problems. What was his score for the test if the teacher gives 100 points for solving the test without mistakes?

3. Find the average of these numbers: 28, -14, -32, 22, -15, 5.

4. Look for a pattern. Give the next three numbers.
   4, 1, -2, -5, ___, ___, ___

5. The sum of a number and (-5) is 23. Find the number.

6. The product of a number and (-4) is 28. Find the number.

7. Find two numbers whose product is 36 and whose sum is -15.

8. One day the temperature in Saint Paul was 7 C. The next day it was -12 C. How many degrees had the temperature dropped?

9. The temperature is rising at the rate of 4 F per hour. In how many hours will it reach 12 F if now it is -8 F ?

10. When an integer is divided by -7, the quotient is -8. Find the integer.

Chapter 10 Test

Order the numbers from greatest to least.

1. -4, 7, | -16 |, -1, 4

2. 22, -33, -11, | -44 |, 2

Add or subtract.

3. -12 + 18 =

4. -16 + (-27) =

5. 13 - 42 =

6. -24 - 38 =

7. 30 - (- 45) =

8. 13 + (-47) =

Multiply or divide.

9. -12 x 5 =

10. -60 ÷ (-5) =

11. 26 x (-7) =

12. -18 x (-18) =

13. -120 ÷ 30 =

14. 483 ÷ (-23) =

Evaluate.

15. 3 x [12 - (-8)] =

16. 30 - 3 x (15 - 18) =

17. 5 x (-2) + (-9) ÷ 3 =

18. -25 - (16 - 9) =

Word problems.

19. The temperature is rising at the rate of 3 F per hour. When will it reach 10 F, if at 6 in the morning it was -5 F?

20. The difference of a number and 18 is (-6). Find the number.

21. Find the average of these numbers: 13, -15, 18, -6, -20

Chapter 1

Lesson 1 Homework 1

1. 600 2. 2,000 3. 5,600 4. 5,400 5. 2,400 6. 200 7. 5,600

8. 810 9. 200 10. 2,400 11. 120 12. 4,800 13. 240,000

14. 16,000 15. 2,700 16. 49,000 17. 200 18. 2,400

19. 18,000 20. 24,000

Lesson 2 Homework 1

1. 48 2. 96 3. 56 4. 50 5. 205 6. 88 7. 215

8. 306 9. 252 10. 90 11. 370 12. 368 13. 408 14. 259

15. 448 16. 581 17. 360 18. 295 19. 864 20. 116 21. 488

22. 609 23. 396 24. 348 25. 57

Lesson 2 Homework 2

1. 140 2. 252 3. 270 4. 576 5. 729 6. 456 7. 204

8. 420 9. 465 10. 144 11. 98 12. 416 13. 136 14. 189

15. 154 16. 372 17. 420 18. 720 19. 168 20. 292 21. 196

22. 118 23. 0 24. 568 25. 480

Lesson 3 Homework 1

1. 1,780 2. 2,376 3. 4,284 4. 3,672 5. 1,722 6. 1,479 7. 1,120

8. 1,050 9. 7,092 10. 612 11. 6,412 12. 1,365 13. 984 14. 2,832

15. 4,208 16. 7,830 17. 676 18. 4,482 19. 2,856 20. 4,466 21. 630

22. 6,912 23. 1,071 24. 5,430 25. 1,548

Lesson 3 Homework 2

1. 1,652 2. 7,578 3. 3,402 4. 3,884 5. 2,421 6. 1,186 7. 2,280

8. 3,045 9. 1,557 10. 2,562 11. 1,431 12. 4,074 13. 5,832 14. 2,564

15. 4,035 16. 1,985 17. 2,466 18. 4,893 19. 6,372 20. 1,629

Lesson 3 Homework 3

1. 938 2. 7,068 3. 276,056 4. 33,648 5. 267,480 6. 37,340

7. 188,031 8. 2,664,487 9. 16,150 10. 23,017,710 11. 454,984,168

12. 5,344,029,192 13. 283,462,928 14. 417,931,584 15. 85,323,987

Lesson 3 Homework 4

1. 2,208 2. 7,496 3. 13,536 4. 7,710 5. 228,956 6. 327,055

7. 250,218 8. 1,512,492 9. 4,876,016 10. 22,249,338 11. 28,842,723

12. 9,876,536 13. 1,830,650,463 14. 1,556,437,300 15. 586,891,782

Lesson 4 Homework 1

1. 18 2. 19 3. 47 4. 13 5. 14 6. 13

7. 12 8. 11 9. 27 10. 23 11. 12 12. 16

Lesson 4 Homework 2

1. 34 2. 15 3. 26 4. 14 5. 14 6. 11

7. 14 8. 10 9. 18 10. 12R3 11. 11R5 12. 15R1

13. 19R1 14. 19R1 15. 11R5

Lesson 5 Homework 1

1. 45 2. 131 3. 55 4. 388 5. 139 6. 29

7. 105 8. 42 9. 136 10. 166 11. 53 12. 116R3

13. 268R1 14. 143R2 15. 156R1

Lesson 5 Homework 2

1. 644R2 2. 9,279 3. 130,575R5 4. 51,322R4 5. 70,900R5

6. 1,446R8 7. 15,034R1 8. 16,707,433 9. 45,517R2

10. 7,084,310R2

Lesson 5 Homework 3

1. 17,805R1 2. 150,690 R2 3. 195,808,766 4. 13,509,065 5. 205,761,315

6. 12,809,029 R5 7. 304,307,765 R6 8. 5,186,452,786 R3 9. 1,096,935,728

10. 609,795,844 R3

Lesson 6 Homework 1

1. 1120 2. 2,220 3. 1,520 4. 2,208 5. 2,016 6. 3,402

7. 4,455 8. 893 9. 1,944 10. 7,852 11. 7,585 12. 45,676

13. 8,773 14. 39,816 15. 34,818

Lesson 6 Homework 2

1. 3,900 2. 3,900 3. 6,510 4. 952 5. 2,756 6. 3,216

7. 17,700 8. 47,040 9. 74,080 10. 17,710 11. 24,516 12. 16,512

13. 38,752 14. 60,048 15. 51,750

Lesson 6 Homework 3

1. 1,624 2. 3,332 3. 6,696 4. 16,110 5. 12,663 6. 21,957

7. 151,767 8. 254,040 9. 491,338 10. 2,524,079 11. 5,152,434

12. 6,571,572 13. 58,274,366 14. 20,744,256 15. 36,183,106

Lesson 7 Homework 1

1. 15,200 2. 136,800 3. 1,144,800 4. 27,135 5. 146,878

6. 159,068 7. 270,600 8. 764,712 9. 2,794,080 10. 3,531,718

11. 3,461,328 12. 670,289 13. 6,312,744 14. 297,917,388 15. 198,266,255

Lesson 7 Homework 2

1. 141,400 2. 378,900 3. 97,500 4. 35,310 5. 359,936

6. 149,275 7. 621,296 8. 2,433,977 9. 4,024,376 10. 27,255,060

11. 45,353,063 12. 41,516,544 13. 75,039,902 14. 306,206,940 15. 515,653,908

Lesson 8 Homework 1

1. 24 2. 32 3. 32 4. 53 5. 43 6. 17

7. 36 8. 24 9. 35 10. 58 11. 37 12. 28

13. 41 14. 48 15. 67

Lesson 8 Homework 2

1. 35 2. 53 3. 61 4. 28 5. 32 6. 57

7. 25R4 8. 33R18 9. 27R8 10. 53R6 11. 44R32 12. 92R8

13. 38R4 14. 15R24 15. 29R41

Lesson 8 Homework 3

1. 134 2. 113 3. 88 4. 214 5. 213 6. 302

7. 123R9 8. 210R8 9. 236R9 10. 201 11. 225R12 12. 207R2

13. 147R8 14. 241R28 15. 333

Lesson 9 Homework 1

1. 6 2. 10 3. 43 4. 5 5. 12 6. 45

7. 140 8. 140 9. 20 10. 315 11. 88% 12. 80 pages

13. $ 27,000 14. 275miles

Lesson 10 Homework 1

1. 2,184 hours 2. 24 candies 3. 525 A's and 350 B's 4. 218,000 people

5. $1,400; $1,300; $1,200 6. Tim has 180 cards; Tom has 140 cards

7. 281,520 min or 4,692 hours 8. $ 1,989 9. 1440 min in one day; 44,640 min in one month; 525,600 min in one year 10. $ 24

Chapter 1 Test

1. 196 2. 2,536 3. 2,772 4. 5,136 5. 3,268 6. 32,188

7. 19,434 8. 1,330,488 9. 37 10. 19 11. 153 12. 77

13. 34 14. 34 15. 26 16. 56

Word problems:

1. 137 apples 2. 1,384 lessons 3. 1,110 calories

Chapter 2

Lesson 1 Homework 1

1. 24 2. 60 3. 54 4. 384 5. 12 6. 4 7. 46 8. 20 9. 20 10. 16

11. 2 12. 0 13. 20 14. 100 15. 2

Lesson 1 Homework 2

1. 32 2. 2 3. 16 4. 22 5. 1 6. 28 7. 20 8. 3 9. 36 10. 13

11. 2 12. 6 13. 10 14. 291 15. 1

Lesson 2 Homework 1

1. 30 2. 30 3. 270 4. 200 5. 2,700 6. 3,400 7. 8,000 8. 381,000

9. 39,000 10. 20,000 11. 4,560,000 12. 500,000 13. 78,800,000

14. 2,300,000 15. 25,000,000 16. 661,000,000 17. 340,000,000 18. 2,880,000,000

Lesson 2 Homework 2

1. 160 2. 90 3. 2,310 4. 4,400 5. 12,300 6. 1,000 7. 70,000

8. 340,000 9. 2,300,000 10. 67,600,000,000 11. 4,400,000,000

12. 24,000,000,000 13. 750,000,000,000 14. 50,000 15. 300,000,000,000

16. 234,000,000 17. 90,210,000,000 18. 3,500,000 19. 200,000,000,000

Lesson 3 Homework 1

1. 80 2. 300 3. 160 4. 44,000 5. 420,000 6. 32,000,000 7. 21,000,000

8. 9,000 9. 16,000 10. 300,000 11. 6,000 12. 32,000,000 13. 400 14. 5

15. 1,900 16. 50 17. 1,000,000

Lesson 3 Homework 2

1. 900 2. 800 3. 1,000 4. 100,000 5. 160,000 6. 6,000,000,000

7. 12,000,000,000 8. 4,200 9. 80,000 10. 720,000 11. 600,000

12. 60,000,000 13. 1,000 14. 3,000,000 15. 10 16. 294,000 17. 30

18. 4,900,000

Lesson 4 Homework 1

1. 800 2. 8,000 3. 13,000 4. 500 5. 7,000 6. 190 7. 530

8. a) 67 b) 24 c) 38 9. a) 8, 42 b) 19, 73 c) 86, 55

Lesson 5 Homework 1

1. (56+10)-1=65 2. (64-10)+2=56 3. (50+26)-2=74 4. (76-20)+1=57

5. (100+16)-1=115 6. (67-50)+2=19 7. (253+400)-2=651 8. (638-200)+1=439

9. (677+100)-3=774 10. (248-100)+3=151 11. (458+200)-1=657

12. (555-300)+2=257 13. (1743+300)-3=2040 14. (2,064-100)+4=1968

15. (6,246+2000)-4=4,242 16. (3,741-3,000)+5=746 17. (40x9)-9=351

18. (45x10)+45=495 19. (78x100)-78=7722 20. (68x1,000)-68=67,932

Lesson 6 Homework 1

1. 30+70=100 2. 73-23=50 3. 170+130=300 4. 390-190=200 5. 20x6=120

6. 80x10=800 7. 30+50+50=130 8. 150-20-30=100 9. 90+21=111

10. 200+24=224 11. 140+49=189 12. 180+24=204 13. 300+30=330

14. 200+40+8=248 15. 80+64=144 16. 720+9=729 17. 900+60+3=963

18. 500+20=520 19. 320+8=328 20. 420+21=441

Lesson 7 Homework 1

1. ≈$1,900 2. $196 3. ≈52 h. 4. Yes 5. ≈210 miles 6. ≈9 h. 7. ≈$2,000

8. ≈$1,250 9. $80 10. ≈40

Chapter 2 Test

1. 10 2. 11 3. 40 4. 40 5. 10 6. 100 7. 150; 80; 280

8. 23,000; 167,000 9. 700,000,000 10. 397,500,000 11. ≈1,600

12. 100+37-1=136 13. ≈900,000 14. ≈ 400x20=8,000 ($) 15. ≈60 mi/h

Chapter 3

Lesson 1 Homework 1

1. Yes 2. No 3. Yes 4. Yes 5. Yes 6. No 7. 1, 2, 3, 4, 6, 12

8. 1, 2, 3, 6, 9, 18 9. 1, 2, 4, 5, 10, 20 10. 1, 2, 4, 7, 14, 28 11. 1, 2, 4, 8, 16, 32

12. 1, 2, 3, 4, 6, 9, 12, 18, 36 13. 1, 29 14. 1, 3, 5, 9, 15, 45

15. 1, 2, 3, 4, 6, 8, 12, 16, 24, 32, 48, 96 16. 4, 8, 12, 16, 20 17. 5, 10, 15, 20, 25

18. 9, 18, 27, 36, 45 19. 12, 24, 36, 48, 60 20. 20, 40, 60, 80, 100

21. 25, 50, 75, 100, 125 22. F 23. T 24. F 25. T 26. T 27. F 28. T

Lesson 1 Homework 2

1. False 2. True 3. False 4. True 5. True 6. False 7. 1,3,5,15

8. 1, 2, 3, 4, 6, 8, 12, 24 9. 1, 2, 4, 5, 8, 10, 20, 40 10. 1, 3, 9, 27 11. 1, 31

12. 1, 2, 4, 7, 8, 14, 28, 56 13. 1, 2, 3, 4, 5, 6, 10, 12, 15, 20, 30, 60

14. 1, 2, 4, 8, 16, 32, 64 15. 1, 2, 3, 4, 6, 8, 9, 12, 18, 24, 36, 72 16. 2, 4, 6, 8, 10

17. 6, 12, 18, 24, 30 18. 8, 16, 24, 32, 40 19. 13, 26, 39, 52, 65

20. 18, 36, 54, 72, 90 21. 30, 60, 90, 120, 150 22. Yes 23. No 24. No

25. Yes 26. Yes 27. No 28. Yes

Lesson 2 Homework 1

1. Yes 2. Yes 3. No 4. Yes 5. Yes 6. No 7. Yes 8. Yes 9. Yes 10. Yes

11. No 12. Yes 13. Yes 14. Yes 15. Yes 16. No 17. Yes 18. No 19. Yes

20. Yes 21. Yes 22. Yes 23. No 24. Yes 25. No 26. Yes 27. Yes 28. No

Lesson 2 Homework 2

1. Yes 2. No 3. Yes 4. No 5. Yes 6. Yes 7. No 8. Yes 9. Yes 10.Yes

11. Yes 12. No 13. Yes 14. Yes 15. No 16. Yes 17. No 18. Yes 19.Yes

20. Yes 21. Yes 22. No 23. Yes 24. Yes 25. Yes 26. No 27. Yes 28.Yes

Lesson 3 Homework 1

1. 1, 2, 3, 6, 9, 18 – composite 2. 1, 19 – prime 3. 1, 2, 4, 8, 16, 32 – composite

4. 1, 37 – prime 5. 1, 2, 3, 4, 6, 8, 12, 16, 24, 48 – composite

6. 1, 2, 3, 4, 5, 6, 10, 12, 15, 20, 30, 60 – composite 7. prime 8. composite: 4×2

9. prime 10. composite: 8×2 11. composite: 6×4 12. composite: 3×9 13. prime

14. composite: 7×7 15. composite: 17×3 16. composite: 3×19 17. composite: 21×3

18. prime 19. 2, 3, 5, 7, 11, 13, 17, 19, 23, 29, 31, 37, 41, 43, 47, 53, 59, 61, 67, 71, 73, 79, 83, 89, 97

Lesson 4 Homework 1

1. 3×4 2. 1×13 3. 5×5 4 5×6 5. 6×7 6. 47×1

7. 18
/\\
9×2
/\\ \\
3×3 ×2

8. 24
/ \\
6 × 4
/ \\ /\\
2×3×2×2

9. 32
/ \\
4 × 8
/\\ /\\
2×2×2×4
| | | /\\
2x2x 2x2x2

10. 36
/ \\
4 × 9
/\\ /\\
2×2×3×3

11. 50
/ \\
25×2
/\\ \\
5×5×2

12. 78
/ \\
2 ×39
/ / \\
2× 3×13

13. 2 | 48
2 | 24
2 | 12
2 | 6
3 | 3
1

14. 3 | 81
3 | 27
3 | 9
3 | 3
1

15. 2 | 96
2 | 48
2 | 24
2 | 12
2 | 6
3 | 3
1

16. 2 | 144
2 | 72
2 | 36
2 | 18
3 | 9
3 | 3
1

```
17. 5 | 225        18. 2 | 98       19. 2 | 120        20. 2 | 250
     5 | 45             7 | 49           2 | 60             5 | 125
       3 | 9              7 | 7            2 | 30             5 | 25
         3 | 3              1                3 | 15             5 | 5
             1                                 5 | 5              1
                                                   1
```

Lesson 4

Homework 2

1. 3×5
2. 3×8
3. 1×31
4. 5×7
5. 1×41
6. 7×8

```
7.    20        8.   36          9.    90         10.  42         11.  96           12.     72
      /\             / \               / \             / \             / \                  /    \
     4×5           9 ×  4            6 × 15           6 × 7          16× 6                 6 ×   12
    /\   \        / \   /\          /\     /\        /\     \        /\    /\             /\     / \
   2×2 ×5        3×3×2×2           2×3×3×5           2×3× 7        4 ×  4×2×3           2×3×3×  4
                                                                  / \   /\  \  \        \  \  \   /\
                                                                2×2×2×2×2×3            2×3×3×2×2
```

```
13. 2 | 54          14. 2 | 250         15. 2 | 48          16. 2 | 100
      3 | 27              5 | 125             2 | 24              2 | 50
        3 | 9               5 | 25              2 | 12              5 | 25
          3 | 3               5 | 5               2 | 6               5 | 5
              1                   1                 3 | 3                 1
                                                        1
```

```
17. 2 | 108         18. 2 | 32          19. 2 | 54          20. 2 | 216
      2 | 54              2 | 16              3 | 27              2 | 108
        3 | 27              2 | 8               3 | 9               2 | 54
          3 | 9               2 | 4               3 | 3               3 | 27
            3 | 3               2 | 2               1                   3 | 9
                1                   1                                     3 | 3
                                                                              1
```

Lesson 5

Homework 1

1. 6
2. 8
3. 14
4. 1, 2, 4, 8
1, 2, 5, 10
GCF = 2
5. 1, 3, 5, 15
1, 2, 4, 5, 10, 20
GCF = 5
6. 1, 3, 7, 21
1, 2, 4, 7, 14, 28
GCF = 7

7. 1, 2, 3, 6
   1, 2, 7, 14
   GCF = 2

8. 1, 2, 4, 8, 16
   1, 2, 4, 5, 8, 10, 20, 40
   GCF = 8

9. 1, 2, 3, 6, 9, 18
   1, 2, 3, 4, 6, 8, 12, 24
   GCF = 6

10. 1, 2, 3, 5, 6, 10, 15, 30;
    GCF = 15

11. 1, 2, 7, 14;
    GCF = 2

12. 1, 2, 4, 8, 16, 32;
    GCF = 16

13. 1, 3, 5, 15
    GCF = 5

14. 1, 2, 11, 22
    GCF = 11

15. 1, 3, 9, 27
    GCF = 9

16. $24 = \underline{2} \times \underline{2} \times 2 \times \underline{3}$
    $36 = \underline{2} \times \underline{2} \times \underline{3} \times 3$
    $GCF = 2 \times 2 \times 3 = 12$

17. $50 = 2 \times \underline{5} \times \underline{5}$
    $75 = \underline{5} \times \underline{5} \times 3$
    $GCF = 5 \times 5 = 25$

18. $42 = 2 \times \underline{3} \times \underline{7}$
    $63 = 3 \times \underline{3} \times \underline{7}$
    $GCF = 3 \times 7 = 21$

19. $54 = 2 \times \underline{3} \times \underline{3} \times \underline{3}$
    $81 = 3 \times \underline{3} \times \underline{3} \times \underline{3}$
    $GCF = 3 \times 3 \times 3 = 27$

20. $25 = 5 \times 5$
    $38 = 2 \times 19$
    GCF = 1

21. $90 = \underline{2} \times \underline{3} \times 3 \times \underline{5}$
    $120 = \underline{2} \times 2 \times 2 \times \underline{3} \times \underline{5}$
    $GCF = 2 \times 3 \times 5 = 30$

22. 6 23. 7 24. 15 25. 8 26. 4 27. 11

Lesson 5 Homework 2

1. 2 2. 6 3. 16

4. 1, 2, 4
   1, 2, 3, 6
   GCF = 2

5. 1, 2, 3, 4, 6, 12
   1, 3, 5, 15
   GCF = 3

6. 1, 2, 4, 8, 16
   1, 2, 3, 4, 6, 9, 12, 18,36
   GCF = 4

7. 1, 3, 9
   1, 3, 11, 33
   GCF = 3

8. 1, 2, 4, 5, 10, 20
   1, 2, 3, 5, 6, 10, 15, 30
   GCF = 10

9. 1, 3, 9, 27
   1, 3, 5, 9, 15, 45
   GCF = 9

10. 1, 2, 4, 7, 14, 28; GCF = 4

11. 1, 2, 11, 22; GCF = 11

12. 1, 2, 3, 4, 6, 8, 12, 24; GCF = 12

13. 1, 13; GCF = 13

14. 1, 2, 3, 6, 7, 14, 21, 42; GCF = 6

15. 1, 3, 9, 27; GCF = 27

16. $21 = 3 \times \underline{7}$
    $28 = 2 \times 2 \times \underline{7}$
    GCF = 7

17. $32 = \underline{2} \times \underline{2} \times \underline{2} \times 2 \times 2$
    $56 = \underline{2} \times \underline{2} \times \underline{2} \times 7$
    $GCF = 2 \times 2 \times 2 = 8$

18. $45 = 3 \times \underline{3} \times \underline{5}$
    $105 = \underline{3} \times \underline{5} \times 7$
    $GCF = 3 \times 5 = 15$

19. $144 = \underline{2}\times\underline{2}\times 2\times 2\times\underline{3}\times\underline{3}$
$108 = \underline{2}\times\underline{2}\times\underline{3}\times\underline{3}\times 3$
$GCF = 2\times 2\times 3\times 3 = 36$

20. $56 = \underline{2}\times\underline{2}\times\underline{2}\times\underline{7}$
$168 = \underline{2}\times\underline{2}\times\underline{2}\times 3\times\underline{7}$
$GCF = 2\times 2\times 2\times 7 = 56$

21. $65 = 5\times\underline{13}$
$91 = 7\times\underline{13}$
$GCF = 13$

22. 3 23. 9 24. 5 25. 8 26. 11 27. 13

Lesson 6 Homework 1

1. 7, 14, 21, 28, 35 2. 3, 6, 9, 12, 15 3. 13, 26, 39, 52, 65 4. 9, 18, 27, 36, 45

5. 6, 12, 18, 24
8, 16, 24
LCM = 24

6. 3, 6, 9, 12, 15
5, 10, 15
LCM = 15

7. 12, 24, 36, 48, 60
15, 30, 45, 60
LCM = 60

8. 4, 8, 12, 16, 20
10, 20
LCM = 20

9. 8, 16, 24
24, 48
LCM = 24

10. 7, 14, 21, 28, 35, 42, 49, 56
8, 16, 24, 32, 40, 48, 56
LCM = 56

11. 21 LCM = 21

12. 25, 50, 75 LCM = 75 13. 20, 40 LCM = 40 14. 10, 20, 30 LCM = 30

15. 12, 24, 36 LCM = 36 16. 18, 36, 54, 72 LCM = 72

17. $12 = 2\times 2\times\underline{3}$
$15 = \underline{3}\times 5$
$LCM = 2\times 2\times 3\times 5 = 60$

18. $8 = \underline{2}\times 2\times 2$
$18 = \underline{2}\times 3\times 3$
$LCM = 2\times 2\times 2\times 3\times 3 = 72$

19. $20 = 2\times 2\times\underline{5}$
$25 = 5\times\underline{5}$
$LCM = 2\times 2\times 5\times 5 = 100$

20. $16 = \underline{2}\times\underline{2}\times\underline{2}\times 2$
$24 = \underline{2}\times\underline{2}\times\underline{2}\times 3$
$LCM=2\times 2\times 2\times 2\times 3=48$

21. $42 = \underline{2}\times 3\times 7$
$70 = \underline{2}\times 5\times 7$
$LCM=2\times 3\times 5\times 7=210$

22. $60 = \underline{2}\times\underline{2}\times\underline{3}\times 5$
$72 = \underline{2}\times\underline{2}\times 2\times\underline{3}\times 3$
$LCM=2\times 2\times 3\times 5\times 2\times 3=360$

23. $30 = \underline{2}\times\underline{3}\times 5$
$36 = \underline{2}\times 2\times\underline{3}\times 3$
$LCM=2\times 3\times 5\times 2\times 3=180$

24. $56 = 2\times 2\times 2\times\underline{7}$
$63 = 3\times 3\times\underline{7}$
$LCM=2\times 2\times 2\times 7\times 3\times 3=504$

25. $22 = 2\times\underline{11}$
$55 = 5\times\underline{11}$
$LCM=2\times 11\times 5=110$

26. 90 27. 504 28. 560 29. 24 30. 90 31. 120 32. 60 33. 36 34. 60

Lesson 6 Homework 2

1. 5, 10, 15, 20, 25 2. 4, 8, 12, 16, 20 3. 10, 20, 30, 40, 50 4. 21, 42, 63, 84, 105

5. 4, 8, 12
6, 12
LCM = 12

6. 2, 4, 6, 8, 10, 12,14
7, 14
LCM = 14

7. 5, 10, 15, 20, 25, 30, 35, 40, 45
9, 18, 27, 36, 45
LCM = 45

8. 12, 24, 36, 48, 60
   15, 30, 45, 60
   LCM = 60

9. 6, 12, 18, 24
   8, 16, 24
   LCM = 24

10. 12, 24, 36
    18, 36
    LCM = 36

11. 21, 42, 63; LCM=63

12. 10, 20; LCM = 20   13. 16, 32, 48; LCM = 48   14. 20, 40, 60, 80; LCM = 80

15. 36, 72, 108; LCM = 108   16. 60, 120, 180; LCM = 180

17. $12 = 2\times2\times3$
    $16 = 2\times2\times2\times2$
    $LCM = 2\times2\times2\times2\times3 = 48$

18. $25 = 5\times5$
    $30 = 5\times2\times3$
    $LCM = 5\times5\times2\times3 = 150$

19. $14 = 7\times2$
    $21 = 7\times3$
    $LCM = 7\times2\times3 = 42$

20. $18 = 2\times3\times3$
    $21 = 3\times7$
    $LCM = 2\times3\times3\times7 = 126$

21. $42 = 2\times3\times7$
    $48 = 2\times3\times2\times2\times2$
    $LCM = 2\times2\times2\times2\times3\times7 = 336$

22. $36 = 2\times2\times3\times3$
    $60 = 2\times2\times3\times5$
    $LCM = 2\times2\times3\times3\times5 = 180$

23. $35 = 5\times7$
    $56 = 2\times2\times2\times7$
    $LCM = 2\times2\times2\times5\times7 = 280$

24. $45 = 3\times3\times5$
    $60 = 2\times2\times3\times5$
    $LCM = 2\times2\times3\times3\times5=180$

25. $33 = 3\times11$
    $44 = 2\times2\times11$
    $LCM = 2\times2\times3\times11 = 132$

26. LCM = 121   27. LCM = 504   28. LCM = 315   29. LCM = 120   30. LCM = 42

31. LCM = 60   32. LCM = 150   33. LCM = 72   34. LCM = 480

Lesson 7 Homework 1

1. February 24   2. 6, 4   3. 3, 60   4. 12 meters   5. April 5   6. 8 pm

7. Every 2 hours   8. 6 sec.   9. 6 times   10. LCM(5,9)

Chapter 3 Test

1. 1, 2, 3, 4, 6, 12 2. 1, 2, 4, 8, 16, 32 3. 1, 2, 4, 5, 10, 20, 25, 50, 100 4. 7, 14, 21, 28, 35 5. 15, 30, 45, 60, 75, 6. 1. 1, 2, 3, 4, 6, 12 2. 1, 2, 4, 8, 16, 32 3. 1, 2, 4, 5, 10, 20, 25, 50, 100 4. 7, 14, 21, 28, 35 5. 15, 30, 45, 60, 75, 6. 21, 42, 63, 84, 105

```
7.        24
         /  \
        6 x 4
       / \   / \
    2 x 3 x 2 x 2

8.        60
         /  \
       4 x 15
      / \   / \
    2 x 2 x 3 x 5

9.        81
         /  \
        9 x 9
       / \   / \
    3 x 3 x 3 x 3
```

```
10. 2 | 80
     2 | 40
      2 | 20
       2 | 10
        5 | 5
          | 1

11. 2 | 144
     2 | 72
      2 | 36
       2 | 18
        3 | 9
         3 | 3
           | 1

12. 2 | 216
     2 | 108
      2 | 54
       3 | 27
        3 | 9
         3 | 3
           | 1
```

13. 6 14. 7 15. 8 16. 72 17. 60 18. 150 19. 3, 12

20. 8,12 21. May 29

Chapter 4

Lesson 1 Homework 1

1. $\frac{1}{2}$, one half 2. $\frac{3}{8}$, three eighths 3. $\frac{3}{4}$, three fourths 4. $\frac{2}{6}$, two sixths

5. $\frac{7}{8}$, seven eighths 6. $\frac{2}{3}$, two thirds 7. $\frac{1}{4}$ 8. $\frac{2}{3}$ 9. $\frac{4}{5}$ 10. $\frac{7}{8}$

11.

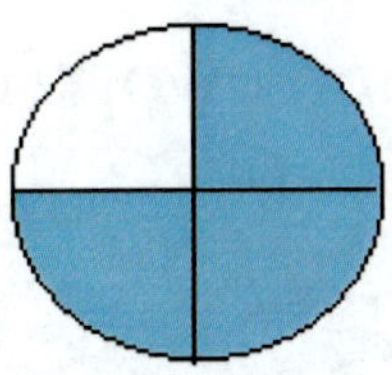

12.

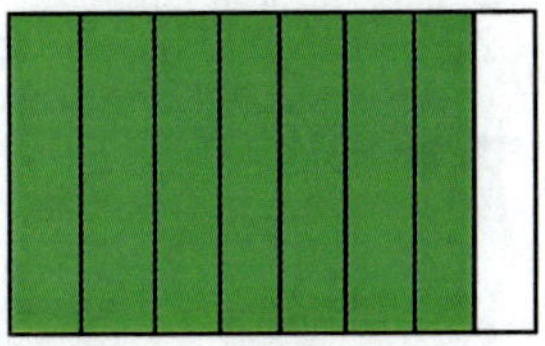

Lesson 1 Homework 2

1. $\frac{1}{2}$, one half 2. $\frac{3}{4}$, three fourths 3. $\frac{1}{4}$, one fourth 4. $\frac{2}{5}$, two fifths

5. $\frac{7}{12}$, seven twelfths 6. $\frac{2}{3}$, two thirds 7. $\frac{3}{7}$ 8. $\frac{5}{10}$ 9. $\frac{10}{11}$ 10. $\frac{4}{13}$

11.

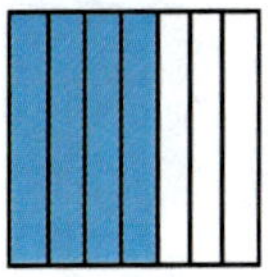

12.

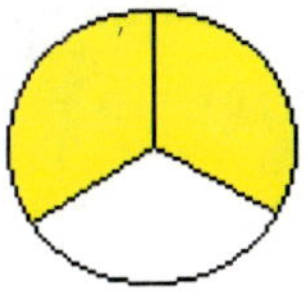

Lesson 2 Homework 1

1. $\frac{2}{3}$ 2. $\frac{5}{7}$ 3. $\frac{3}{5}$ 4. $\frac{7}{10}$ 5. 1) $\frac{4}{12}$ 2) $\frac{7}{12}$ 6. $\frac{2}{7}$ 7. $\frac{1}{2}$

8. seven eighths 9. three tenths 10. six thirteenths 11. 7 12. 12 13. 35

Lesson 2 Homework 2

1. $\frac{3}{4}$ 2. $\frac{4}{7}$ 3. $\frac{4}{6}$ 4. $\frac{5}{5}$ 5. 1) $\frac{6}{15}$ 2) $\frac{5}{15}$ 3) $\frac{4}{15}$

6.

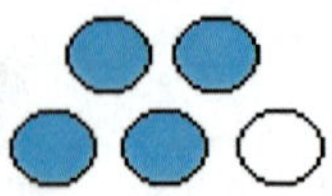

7.

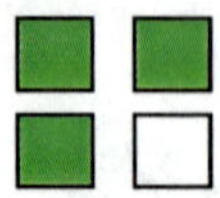

8.

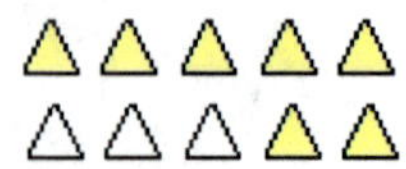

9. 3 10. 32 11. 15

Lesson 3 Homework 1

1. $\frac{1}{4}=\frac{2}{8}$ 2. $\frac{1}{2}=\frac{2}{4}$ 3. $\frac{2}{3}=\frac{4}{6}$ 4. $\frac{4}{6}$ 5. $\frac{10}{16}$ 6. $\frac{14}{24}$ 7. $\frac{9}{12}$

8. $\frac{12}{21}$ 9. $\frac{6}{15}$ 10. $\frac{6}{10}=\frac{9}{15}=\frac{12}{20}$ 11. $\frac{4}{18}=\frac{6}{27}=\frac{8}{36}$

12. $\frac{8}{14}=\frac{12}{21}=\frac{16}{28}$ 13. $\frac{10}{12}=\frac{15}{18}=\frac{20}{24}$

Lesson 3 Homework 2

1. $\frac{1}{2}=\frac{3}{6}$ 2. $\frac{1}{3}=\frac{2}{6}$ 3. $\frac{1}{4}=\frac{3}{12}$ 4. $\frac{4}{8}$ 5. $\frac{12}{16}$ 6. $\frac{16}{24}$ 7. $\frac{12}{16}$

8. $\frac{16}{28}$ 9. $\frac{8}{20}$ 10. $\frac{2}{6}=\frac{3}{9}=\frac{4}{12}$ 11. $\frac{4}{10}=\frac{6}{15}=\frac{8}{20}$

12. $\frac{10}{14}=\frac{15}{21}=\frac{20}{28}$ 13. $\frac{22}{24}=\frac{33}{36}=\frac{44}{48}$

Lesson 3 Homework 3

1. $\frac{6}{9}$ 2. $\frac{6}{8}$ 3. $\frac{20}{25}$ 4. $\frac{21}{24}$ 5. $\frac{25}{45}$ 6. $\frac{4}{24}$ 7. $\frac{16}{28}$ 8. $\frac{20}{50}$

9. $\frac{45}{55}$ 10. $\frac{66}{72}$ 11. $\frac{64}{120}$ 12. $\frac{90}{99}$ 13. $\frac{72}{84}$ 14. $\frac{49}{84}$ 15. $\frac{7}{56}$

16. $\frac{80}{104}$ 17. $\frac{36}{56}$ 18. $\frac{50}{54}$ 19. $\frac{21}{27}$ 20. $\frac{90}{111}$ 21. $\frac{85}{100}$

Lesson 4 Homework 1

1. Yes 2. No 3. No 4. Yes 5. No 6. Yes 7. Yes 8. No

9. $\frac{1}{4}$ 10. $\frac{1}{3}$ 11. $\frac{1}{3}$ 12. $\frac{2}{3}$ 13. $\frac{4}{5}$ 14. $\frac{3}{4}$ 15. $\frac{1}{3}$ 16. $\frac{1}{3}$

17. $\frac{1}{4}$ 18. $\frac{1}{4}$ 19. $\frac{1}{3}$ 20. $\frac{1}{4}$ 21. $\frac{1}{5}$ 22. $\frac{4}{5}$ 23. $\frac{1}{2}$ 24. $\frac{2}{3}$

25. $\frac{5}{6}$ 26. $\frac{2}{3}$ 27. $\frac{1}{5}$ 28. $\frac{1}{4}$ 29. $\frac{1}{5}$

Lesson 4 Homework 2

1. No 2. Yes 3. No 4. Yes 5. Yes 6. No 7. Yes 8. No

9. $\frac{1}{2}$ 10. $\frac{3}{7}$ 11. $\frac{8}{15}$ 12. $\frac{7}{12}$ 13. $\frac{1}{7}$ 14. $\frac{2}{5}$ 15. $\frac{2}{5}$ 16. $\frac{1}{3}$

17. $\frac{25}{28}$ 18. $\frac{7}{9}$ 19. $\frac{16}{21}$ 20. $\frac{3}{4}$ 21. $\frac{5}{7}$ 22. $\frac{6}{7}$ 23. $\frac{2}{9}$ 24. $\frac{5}{6}$ 25. $\frac{6}{7}$ 26. $\frac{32}{45}$

Lesson 4 Homework 3

1. $\frac{1}{3}$ 2. $\frac{1}{6}$ 3. $\frac{1}{4}$ 4. $\frac{1}{9}$ 5. $\frac{1}{16}$ 6. $\frac{2}{7}$ 7. $\frac{1}{6}$ 8. $\frac{1}{4}$ 9. $\frac{20}{33}$

10. $\frac{9}{16}$ 11. $\frac{2}{3}$ 12. $\frac{1}{7}$ 13. $\frac{4}{5}$ 14. $\frac{5}{7}$ 15. $\frac{7}{8}$ 16. $\frac{1}{25}$ 17. $\frac{1}{50}$

18. $\frac{1}{100}$ 19. $\frac{1}{20}$ 20. $\frac{1}{8}$ 21. $\frac{1}{30}$ 22. $\frac{1}{5}$ 23. $\frac{9}{10}$ 24. $\frac{7}{8}$

25. $\frac{1}{11}$ 26. $\frac{1}{12}$ 27. $\frac{89}{90}$

Lesson 5 Homework 1

1. $\frac{19}{3}$ 2. $\frac{29}{5}$ 3. $\frac{58}{7}$ 4. $\frac{37}{5}$ 5. $\frac{68}{7}$ 6. $\frac{43}{9}$ 7. $\frac{80}{7}$ 8. $\frac{97}{7}$

9. $\frac{109}{10}$ 10. $\frac{104}{5}$ 11. $\frac{65}{3}$ 12. $\frac{75}{4}$ 13. $3\frac{2}{7}$ 14. $3\frac{3}{9}$ 15. $6\frac{1}{3}$

16. $4\frac{4}{9}$ 17. $6\frac{1}{8}$ 18. $14\frac{2}{5}$ 19. $9\frac{4}{7}$ 20. $13\frac{3}{4}$ 21. $9\frac{1}{11}$ 22. $9\frac{7}{9}$

23. $8\frac{3}{7}$ 24. $11\frac{7}{8}$

Lesson 5 Homework 2

1. $\frac{11}{3}$ 2. $\frac{47}{8}$ 3. $\frac{39}{4}$ 4. $\frac{53}{7}$ 5. $\frac{29}{6}$ 6. $\frac{42}{5}$ 7. $\frac{53}{5}$ 8. $\frac{37}{3}$

9. $\frac{83}{7}$ 10. $\frac{103}{6}$ 11. $\frac{104}{5}$ 12. $\frac{103}{4}$ 13. $2\frac{1}{7}$ 14. $2\frac{2}{9}$ 15. $4\frac{4}{5}$

16. $7\frac{1}{5}$ 17. $5\frac{5}{8}$ 18. $7\frac{1}{7}$ 19. $9\frac{3}{5}$ 20. $6\frac{1}{10}$ 21. $4\frac{5}{6}$ 22. $7\frac{9}{13}$

23. $7\frac{4}{11}$ 24. $11\frac{1}{4}$

Lesson 6 Homework 1

1. 9 students 2. 3 apples 3. 75 pages 4. Max ate 200gr; Sam ate 300gr

5. 345 miles 6. $\frac{5}{10}$ 7. 1 hour 8. $\frac{3}{4}$ 9. Yes 10. 7 students

Chapter 4 Test

1. $\frac{3}{4}$ 2. $\frac{5}{8}$ 3.

4. a) $\frac{5}{10} = \frac{1}{2}$ b) $\frac{7}{10}$ 5. a) $\frac{12}{16}$ b) $\frac{5}{8}$ c) $\frac{16}{56}$ 6. a) $\frac{1}{4}$ b) $\frac{2}{3}$ c) $\frac{1}{3}$

7. $\frac{5}{8}, \frac{1}{2}, \frac{2}{5}, \frac{3}{8}, \frac{1}{3}$ 8. a) $\frac{38}{7}$ b) $\frac{47}{5}$ c) $\frac{61}{4}$ 9. 15 flowers

Chapter 5

Lesson 1 — Homework 1

1. $\frac{4}{5}$ 2. $\frac{1}{4}$ 3. $\frac{1}{2}$ 4. $\frac{1}{2}$ 5. $\frac{2}{3}$ 6. $\frac{3}{4}$ 7. 1 8. $\frac{4}{11}$ 9. $1\frac{1}{7}$

10. $\frac{1}{4}$ 11. $1\frac{1}{5}$ 12. $1\frac{1}{3}$ 13. 1 14. $\frac{2}{3}$ 15. $1\frac{2}{5}$ 16. $1\frac{1}{8}$

Lesson 2 — Homework 1

1. $\frac{7}{9}$ 2. $\frac{1}{4}$ 3. $\frac{7}{10}$ 4. $\frac{1}{3}$ 5. $\frac{3}{4}$ 6. $\frac{1}{12}$ 7. $1\frac{5}{24}$ 8. $\frac{7}{12}$

9. $1\frac{3}{10}$ 10. $\frac{11}{24}$ 11. $1\frac{4}{15}$ 12. $\frac{5}{21}$ 13. $\frac{35}{36}$ 14. $\frac{1}{18}$ 15. $1\frac{1}{6}$

16. $\frac{11}{42}$ 17. $1\frac{1}{4}$ 18. $\frac{17}{36}$ 19. $1\frac{11}{40}$ 20. $\frac{2}{9}$

Lesson 2 — Homework 2

1. $1\frac{1}{4}$ 2. $\frac{1}{6}$ 3. $\frac{10}{21}$ 4. $\frac{1}{12}$ 5. $1\frac{19}{30}$ 6. $\frac{3}{8}$ 7. $\frac{13}{24}$ 8. $\frac{23}{42}$

9. $1\frac{13}{18}$ 10. $\frac{13}{24}$ 11. $1\frac{13}{40}$ 12. $\frac{13}{30}$ 13. $1\frac{17}{60}$ 14. $\frac{23}{66}$ 15. $1\frac{17}{36}$

16. $\frac{19}{44}$ 17. $1\frac{7}{26}$ 18. $\frac{8}{45}$ 19. $1\frac{5}{24}$ 20. $\frac{1}{6}$

Lesson 3 — Homework 1

1. $1\frac{3}{7}$ 2. $4\frac{5}{7}$ 3. $10\frac{9}{10}$ 4. 8 5. $4\frac{2}{3}$ 6. $9\frac{7}{10}$ 7. $8\frac{6}{11}$ 8. $9\frac{10}{13}$

9. $13\frac{6}{7}$ 10. $8\frac{1}{4}$ 11. $19\frac{2}{3}$ 12. $16\frac{1}{5}$ 13. $21\frac{2}{7}$ 14. $32\frac{2}{13}$ 15. 16

16. $19\frac{2}{11}$ 17. 20 18. $29\frac{2}{3}$ 19. 8

Lesson 3 Homework 2

1. $6\frac{4}{5}$ 2. $7\frac{3}{4}$ 3. $13\frac{5}{7}$ 4. $14\frac{1}{4}$ 5. $20\frac{1}{3}$ 6. 31 7. $31\frac{1}{2}$

8. $41\frac{2}{7}$ 9. 62 10. $51\frac{1}{2}$ 11. $35\frac{3}{4}$ 12. $54\frac{6}{7}$ 13. $41\frac{1}{2}$

14. $78\frac{2}{7}$ 15. 100 16. $86\frac{6}{7}$ 17. $116\frac{5}{11}$ 18. $164\frac{3}{13}$ 19. $77\frac{1}{3}$

Lesson 4 Homework 1

1. $3\frac{7}{9}$ 2. $4\frac{1}{2}$ 3. $5\frac{2}{7}$ 4. $30\frac{2}{5}$ 5. $6\frac{1}{2}$ 6. $19\frac{7}{11}$ 7. $9\frac{2}{3}$ 8. $3\frac{5}{7}$

9. $\frac{3}{5}$ 10. $\frac{2}{3}$ 11. $9\frac{1}{2}$ 12. $2\frac{2}{3}$ 13. $6\frac{5}{11}$ 14. $34\frac{1}{2}$ 15. $5\frac{5}{6}$

16. $51\frac{1}{4}$ 17. $19\frac{1}{2}$ 18. $14\frac{3}{5}$ 19. $89\frac{2}{3}$

Lesson 4 Homework 2

1. $6\frac{2}{5}$ 2. $7\frac{3}{7}$ 3. $16\frac{2}{11}$ 4. $20\frac{1}{4}$ 5. $37\frac{1}{3}$ 6. $40\frac{1}{3}$ 7. $5\frac{1}{2}$

8. $19\frac{3}{4}$ 9. $45\frac{2}{5}$ 10. $13\frac{4}{7}$ 11. $3\frac{2}{3}$ 12. $5\frac{1}{2}$ 13. $1\frac{1}{6}$ 14. $7\frac{2}{3}$

15. $23\frac{2}{3}$ 16. $9\frac{9}{13}$ 17. $39\frac{2}{5}$ 18. $70\frac{5}{8}$ 19. $9\frac{6}{7}$

Lesson 5 Homework 1

1. $7\frac{3}{4}$ 2. $4\frac{1}{10}$ 3. $11\frac{5}{8}$ 4. $7\frac{1}{9}$ 5. $12\frac{11}{15}$ 6. $5\frac{7}{30}$ 7. $25\frac{13}{14}$

8. $5\frac{1}{10}$ 9. $10\frac{1}{2}$ 10. $4\frac{3}{4}$ 11. $16\frac{5}{8}$ 12. $\frac{5}{9}$ 13. $21\frac{11}{20}$ 14. $7\frac{5}{7}$

15. $31\frac{13}{30}$ 16. $6\frac{1}{6}$ 17. $21\frac{11}{18}$ 18. $34\frac{1}{12}$ 19. $51\frac{4}{21}$

Lesson 5 Homework 2

1. $12\frac{1}{2}$ 2. $4\frac{1}{2}$ 3. $12\frac{11}{12}$ 4. $5\frac{1}{6}$ 5. $22\frac{17}{24}$ 6. $8\frac{6}{13}$ 7. $22\frac{1}{18}$

8. $7\frac{5}{21}$ 9. $51\frac{7}{60}$ 10. $20\frac{11}{24}$ 11. $36\frac{7}{30}$ 12. $2\frac{9}{28}$ 13. $23\frac{17}{42}$

14. $24\frac{17}{48}$ 15. $48\frac{17}{40}$ 16. $6\frac{1}{10}$ 17. $85\frac{17}{60}$ 18. $\frac{17}{18}$ 19. $105\frac{1}{3}$

Lesson 5 Homework 3

1. $6\frac{1}{6}$ 2. $3\frac{1}{4}$ 3. $10\frac{7}{12}$ 4. $1\frac{14}{15}$ 5. $16\frac{5}{18}$ 6. $1\frac{9}{10}$ 7. $19\frac{17}{56}$

8. $20\frac{7}{24}$ 9. $41\frac{9}{20}$ 10. $21\frac{4}{19}$ 11. $71\frac{25}{36}$ 12. $29\frac{4}{5}$ 13. $23\frac{11}{12}$

14. $20\frac{19}{24}$ 15. $58\frac{29}{30}$ 16. $63\frac{1}{4}$ 17. $16\frac{3}{4}$ 18. $3\frac{1}{6}$ 19. $3\frac{1}{6}$ 20. $14\frac{13}{24}$

Lesson 6 Homework 1

1. $\frac{5}{10} < \frac{6}{10}$ 2. $\frac{6}{8} = \frac{6}{8}$ 3. $\frac{28}{35} > \frac{25}{35}$ 4. $\frac{60}{140} > \frac{56}{140}$ 5. $\frac{7}{10} < \frac{9}{10}$

6. $\frac{18}{24} = \frac{18}{24}$ 7. $\frac{5}{15} > \frac{3}{15}$ 8. $\frac{28}{40} < \frac{35}{40}$ 9. $\frac{35}{42} > \frac{30}{42}$ 10. $\frac{3}{8} > \frac{7}{19}$ $57 > 56$

11. $\frac{9}{27} = \frac{3}{9}$ $81 = 81$ 12. $\frac{11}{12} < \frac{14}{15}$ $165 < 168$ 13. $\frac{15}{17} > \frac{12}{15}$ $225 > 204$ 14. $\frac{20}{21} > \frac{17}{18}$ $360 > 357$ 15. $\frac{7}{9} = \frac{21}{27}$ $189 = 189$

16. $\frac{1}{6}, \frac{1}{4}, \frac{1}{3}, \frac{5}{12}$ 17. $\frac{1}{2}, \frac{3}{5}, \frac{2}{3}, \frac{3}{4}$ 18. $\frac{12}{19}, \frac{2}{3}, \frac{14}{17}, \frac{9}{10}$

Lesson 6 Homework 2

1. $\frac{8}{12} < \frac{9}{12}$ 2. $\frac{8}{14} > \frac{7}{14}$ 3. $\frac{77}{88} < \frac{80}{88}$ 4. $\frac{24}{44} = \frac{24}{44}$ 5. $\frac{80}{110} > \frac{77}{110}$

6. $\frac{72}{90} < \frac{75}{90}$ 7. $\frac{28}{30} > \frac{27}{30}$ 8. $\frac{60}{120} = \frac{60}{120}$ 9. $\frac{11}{13} < \frac{12}{13}$ 10. $\frac{9}{14} > \frac{8}{13}$ $117 > 112$

11. $\frac{22}{23} > \frac{23}{25}$ $550 > 529$ 12. $\frac{16}{24} = \frac{12}{18}$ $288 = 288$ 13. $\frac{11}{20} > \frac{15}{29}$ $319 > 300$ 14. $\frac{18}{20} < \frac{10}{11}$ $198 < 200$ 15. $\frac{25}{100} < \frac{27}{100}$ $2500 < 2700$

16. $\frac{12}{19}, \frac{2}{3}, \frac{14}{17}, \frac{9}{10}$ 17. $\frac{3}{10}, \frac{7}{22}, \frac{1}{3}, \frac{4}{11}$

Lesson 7 Homework 1

1. 6+7=13 2. 12- 3=9 3. $7 + 8\frac{1}{2} + 10 = 25\frac{1}{2}$ 4. 19-7=12 5. 3+4+2+1=10

6. 52-37=15 7. $16\frac{1}{2} + 24\frac{1}{2} = 41$ 8. 49-18=31 9. 25+30+15+1=71

10. $120\frac{1}{2} - 83\frac{1}{2} = 37$ 11. $88\frac{1}{2} + 111\frac{1}{2} = 200$ 12. 69-36=33 13. 22+33+42=97

14. 97-16=81 15. 17+23+30=70 16. 77-17=60 17. $26 + 38\frac{1}{2} + 25\frac{1}{2} = 90$

18. 51-21 = 30 19. 10+20+30 = 60 20. 100-50 = 50

Lesson 8 Homework 1

1. $40\frac{1}{6}$ 2. $\frac{1}{8}$ 3. $4\frac{1}{10}$ 4. $\frac{11}{15}$ 5. $25\frac{11}{12}$ 6. $2\frac{1}{2}$ 7. $12\frac{1}{3}$ 8. $\frac{8}{45}$

9. $\frac{7}{20}$ 10. 5

Chapter 5 Test

1. $\frac{3}{4}$ 2. $\frac{11}{24}$ 3. $\frac{1}{6}$ 4. $\frac{7}{36}$ 5. $1\frac{7}{40}$ 6. $\frac{7}{12}$ 7. $4\frac{8}{13}$ 8. $1\frac{4}{7}$

9. $6\frac{2}{3}$ 10. $7\frac{7}{8}$ 11. $5\frac{1}{10}$ 12. 14 13. $7\frac{5}{6}$ 14. $31\frac{17}{24}$ 15. $20\frac{23}{56}$

16. 8+9=17 17. 18-13=5 18. 60+41=101 19. $9\frac{1}{6}$ meters 20. $\frac{28}{45}$ of a mile

Chapter 6

Lesson 1 Homework 1

1. 9 2. 8 3. 15 4. 15 5. 21 6. 4 7. 36 8. 11 9. 42 10. 85

11. 40 12. 28 13. $7\frac{1}{5}$ 14. $7\frac{7}{9}$ 15. $12\frac{6}{7}$ 16. $8\frac{8}{9}$ 17. $16\frac{4}{5}$

18. $8\frac{2}{11}$ 19. 40 20. 15

Lesson 2 Homework 1

1. $\frac{1}{6}$ 2. $\frac{5}{8}$ 3. $\frac{8}{11}$ 4. $\frac{16}{21}$ 5. $\frac{3}{10}$ 6. $\frac{3}{4}$ 7. $\frac{40}{81}$ 8. $\frac{4}{9}$ 9. $\frac{13}{21}$ 10. $\frac{12}{35}$

11. $\frac{8}{9}$ 12. $\frac{5}{19}$ 13. $\frac{24}{35}$ 14. $\frac{3}{7}$ 15. $\frac{63}{110}$ 16. 10 17. $\frac{16}{45}$ 18. 36

Lesson 2 Homework 2

1. $\frac{2}{3}$ 2. $\frac{14}{33}$ 3. $\frac{1}{12}$ 4. $\frac{14}{27}$ 5. $\frac{15}{28}$ 6. $\frac{21}{40}$ 7. $\frac{1}{6}$ 8. $\frac{9}{28}$ 9. $\frac{17}{39}$ 10. $\frac{3}{5}$

11. $\frac{9}{2}$ 12. $\frac{28}{3}$ 13. $\frac{15}{112}$ 14. $\frac{16}{23}$ 15. $\frac{4}{15}$ 16. $\frac{2}{9}$ 17. $\frac{4}{9}$ 18. $\frac{5}{6}$

Lesson 3 Homework 1

1. 35 2. $13\frac{1}{2}$ 3. $5\frac{4}{7}$ 4. $38\frac{1}{2}$ 5. 21 6. $37\frac{1}{2}$ 7. $17\frac{1}{3}$ 8. $52\frac{1}{2}$ 9. $52\frac{1}{2}$

10. $17\frac{1}{2}$ 11. $41\frac{2}{5}$ 12. $66\frac{1}{2}$ 13. $4\frac{1}{5}$ 14. $6\frac{6}{7}$ 15. $13\frac{1}{2}$ 16. $9\frac{1}{6}$

17. $14\frac{7}{12}$ 18. $\frac{3}{4}$ 19. 35 20. $13\frac{1}{2}$

Lesson 3 Homework 2

1. 62 2. 32 3. 60 4. 90 5. $4\frac{8}{9}$ 6. $2\frac{1}{10}$ 7. 84 8. $8\frac{5}{9}$ 9. $47\frac{1}{2}$

10. $16\frac{1}{3}$ 11. $10\frac{1}{2}$ 12. $51\frac{2}{3}$ 13. 236 14. 148 15. $34\frac{1}{8}$ 16. $9\frac{3}{4}$

17. $32\frac{2}{3}$ 18. $22\frac{1}{2}$ 19. 32 20. $49\frac{1}{3}$

Lesson 4 Homework 1

1. $\frac{3}{1}$ 2. $\frac{9}{4}$ 3. $\frac{2}{7}$ 4. $\frac{1}{5}$ 5. $\frac{4}{11}$ 6. $\frac{5}{18}$ 7. $1\frac{1}{5}$ 8. 2 9. $\frac{13}{14}$ 10. $3\frac{1}{3}$

11. $\frac{6}{7}$ 12. $3\frac{1}{3}$ 13. $1\frac{1}{20}$ 14. $1\frac{1}{9}$ 15. $1\frac{1}{3}$ 16. $1\frac{3}{5}$ 17. $\frac{5}{24}$ 18. $\frac{17}{28}$

19. $4\frac{1}{5}$ 20. $1\frac{2}{3}$

Lesson 4 Homework 2

1. $\frac{5}{2}$ 2. $\frac{3}{7}$ 3. $\frac{11}{4}$ 4. $\frac{1}{12}$ 5. $\frac{7}{39}$ 6. $\frac{9}{73}$ 7. $1\frac{1}{4}$ 8. $\frac{9}{16}$ 9. 3 10. $\frac{5}{6}$

11. $\frac{14}{15}$ 12. $1\frac{17}{27}$ 13. $\frac{2}{5}$ 14. $2\frac{11}{12}$ 15. $\frac{1}{2}$ 16. 4 17. $1\frac{7}{15}$ 18. $1\frac{5}{28}$

19. $1\frac{4}{5}$ 20. $1\frac{1}{36}$

Lesson 5 Homework 1

1. $\frac{1}{2}$ 2. 3 3. $1\frac{1}{2}$ 4. $\frac{2}{3}$ 5. 6 6. $1\frac{5}{7}$ 7. 8 8. $\frac{1}{6}$ 9. 1 10. $\frac{4}{9}$

11. $7\frac{1}{2}$ 12. $5\frac{5}{7}$ 13. $1\frac{5}{7}$ 14. $3\frac{3}{7}$ 15. $2\frac{4}{5}$ 16. $3\frac{3}{4}$ 17. $\frac{10}{11}$ 18. $1\frac{1}{3}$

19. 9 20. $7\frac{1}{2}$

Lesson 5 Homework 2

1. $1\frac{1}{2}$ 2. $4\frac{1}{2}$ 3. $1\frac{1}{48}$ 4. $6\frac{2}{3}$ 5. $\frac{3}{5}$ 6. $\frac{5}{7}$ 7. $\frac{4}{9}$ 8. $2\frac{2}{3}$ 9. $4\frac{1}{2}$

10. $\frac{3}{4}$ 11. $6\frac{1}{14}$ 12. $1\frac{1}{3}$ 13. $2\frac{2}{27}$ 14. $6\frac{2}{3}$ 15. $3\frac{1}{3}$ 16. 6 17. $3\frac{3}{8}$

18. $\frac{21}{22}$ 19. $1\frac{3}{7}$ 20. $4\frac{1}{2}$

Lesson 6 Homework 1

(results can very)

1. 20 2. 56 3. 100 4. 56 5. 48 6. 60 7. 88 8. 36 9. 132 10. 80

11. 40 12. 100 13. 75 14. 70 15. 120 16. 35 17. 90 18. 100

Lesson 6 Homework 2

(results can very)

1. 3 2. 1 3. 3 4. 3 5. 5 6. 5 7. 7 8. 5 9. 7 10. 3 11. 7 12. 10

13. 9 14. 8 15. 4 16. 20 17. 3 18. 1

Lesson 7 Homework 1

1. 6 days 2. $1\frac{1}{2}$ hour 3. 23 miles 4. 80 dollars 5. $\frac{2}{3}$ inches 6. 34 m^2

7. \$275 8. 6 watermelons 9. $27\frac{5}{9}$ hours 10. $12\frac{11}{12}$ m^2

Chapter 6 Test

1. $\frac{1}{8}$ 2. 21 3. 4 4. $\frac{3}{5}$ 5. $\frac{3}{20}$ 6. $2\frac{1}{2}$ 7. 21 8. 16 9. 110 10. $\frac{3}{4}$

11. $1\frac{1}{3}$ 12. 14 13. $\frac{1}{21}$ 14. $\frac{7}{10}$ 15. $\frac{1}{2}$ 16. $1\frac{7}{9}$ 17. $1\frac{1}{9}$ 18. $2\frac{2}{5}$

19. 105 in^2 20. $\frac{3}{4}$ lb

Chapter 7

Lesson 1 Homework 1

1. three tenths 2. seven hundredths 3. five thousandths 4. twenty-five hundredths

5. thirty-eight thousandths 6. two hundred forty-eight thousandths

7.one and four tenths 8. twenty eight and five hundredths

9. ten and six hundred four thousandths 10. 0+4x0.1+3x0.01 11. 3+2x0.1+7x0.001

12. 2x10+4+3x0.1 13. 3x100+2x10+1+1x0.1+2x0.01+3x0.001 14. 6x10+5+7x0.01

15. 0+8x0.1+7x0.01+9x0.001 16. 0.7 17. 0.52 18. 0.382 19. 5.03 20. 12.023

21. 3.6 22. 0.8 23. 2.047 24. 0.153 25. 75.94 26. four tenths

27. five thousandths 28. two hundredths 29. $\frac{3}{10}$ 30. $\frac{84}{100}$ 31. $2\frac{72}{100}$ 32. $17\frac{245}{1000}$

Lesson 1 Homework 2

1. 0.9 2. 0.37 3. 0.457 4. 7.3 5. 28.038 6. 45.47

7. three and five tenths = 3+5x0.1 8. Five hundredths = 0+5x0.01

9. four hundred fifty-three thousandths = 0+4x0.1+5x0.01+3x0.001

10. two hundred four thousandths = 0+2x0.1+4x0.001

11. twenty two and twenty-two hundredths = 2x10+2+2x0.1+2x0.01

12. seventeen and thirty-two thousandths = 1x10 +7+3x0.01+2x0.001

13. 0.07 14.20.369 15. 8.8 16. Three hundredths 17. Five ones

18. nine thousandths 19. 5.997, 23.928 20. 5.997, 6.097

21. 14.328, 23,928, 44.729 22. $\frac{5}{10}$ 23. $3\frac{22}{100}$ 24. $25\frac{75}{1000}$ 25. $\frac{349}{1000}$

Lesson 2 Homework 1

1. 0.0079 2. 0.02345 3. 0.877651 4. 12.0003 5. 5.56789 6. 18.045007

7. seven hundred-thousandths = 0+7x0.00001
8. two and one hundred twenty-three thousand four hundred fifty-six millionths = 2+1x0.1+2x0.01+3x0.001+4x0.0001+5x0.00001+6x0.000001

9. nine and four thousand five hundred five millionths = 9+4x0.001+50.0001+5x0.000001

10. thirty four and one hundred twenty-three millionths 3x10+4+1x0.0001+2x0.00001+3x0.000001

11. twenty-five thousand two hundred fifty-seven hundred-thousandths = 0+2x0.1+5x0.01+2x0.001+5x0.0001+7x0.00001

12. twelve and three thousand four hundred fifty six ten-thousandths = 1x10+2+3x0.1+4x0.01+5x0.001+6x0.0001

13. 0.0045 14. 37.463207 15. 2.03569 16. two millionths

17. three ten-thousandths 18. four hundred 19. six ten-thousandths

20. 0.000153; 0.023983 21. 3.46015; 0.000153

22. 3.46015; 0.000153; 82.22212; 0.35113 23. $\frac{9}{2{,}000}$ 24. $\frac{511}{20{,}000}$ 25. $\frac{1}{3125}$

Lesson 2 Homework 2

1. 0.07834 2. 0.0305 3. 0.234567 4. 6.0059 5. 27.00007 6. 3.000357

7. three and three hundred forty-five hundred-thousandths = 3+3x0.001+4x0.0001+5x0.00001

8. twenty seven and two hundred five thousand six hundred seven millionths = 2x10+7+2x0.1+5x0.001+6x0.0001+7x0.000001

9. three thousand four hundred seven hundred-thousandths = 3x0.01+4x0.001+7x0.00001

10. nine hundred ninety-nine thousand nine hundred ninety-nine millionths = 9x0.1+9x0.01+9x0.001+9x0.0001+9x0.00001+9x0.000001

11. six hundred thirty-five and six thousand eight hundred five ten-thousandths = 6x100+3x10+5+6x0.1+8x0.01+5x0.0001

12. four and four hundred thirty two millionths = 4+4x0.0001+3x0.00001+2x0.000001

13. 0.035770 14. 3.00009 15. 75.3096 16. six ten-thousandths

17. seven millionths 18. four thousandths 19. Five hundred-thousandths

20. 0.234626; 7.020426; 21.22942; 0.00312 21. 21.22942; 7.020426; 5.023456

22. 0.234626; 7.020426; 5.023456 23. $\frac{13}{2500}$ 24. $24\frac{11}{4,000}$ 25. $3\frac{9}{125}$

Lesson 3 Homework 1

1. 0.6 2. 2.17 3. 0.33 4. > 5. < 6. < 7. =

8. 0.0344; 0.0345; 0.0354 9. 2.17809; 2.17869; 2.17888; 2.17896 10. 0.2; 1.9

11-13.

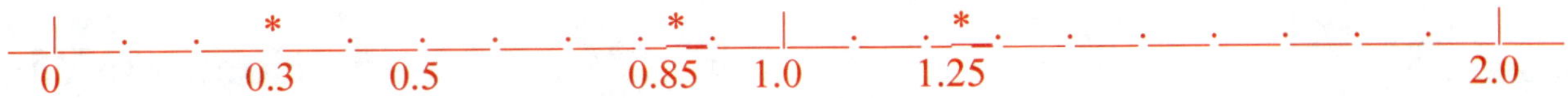

Lesson 3 Homework 2

1. 0.4 2. 3.6 3. 3.33 4. > 5. < 6. < 7. <

8. 8.0; 7.8; 0.87; 0.79 9. 0.0101; 0.0021; 0.0012; 0.00098 10. 7.6; 8.9

11-13.

Lesson 4 Homework 1

1. 1 2. 34 3. 5 4. 7.6 5. 0.1 6. 200.0 7. 0.23 8. 4.23 9. 0.30

10. 12.333 11. 0.005 12. 6.669 13. 0.0035 14. 2.9355 15. 22.1000

16. 5.8 17. 0.3403 18. 0.0025 19. 0.30 20. 25 21. 0.0568 22. 50

23. 2 24. 252 25. 32

Lesson 4 Homework 2

1. 0 2. 6 3. 30 4. 0.8 5. 32.1 6. 20.2 7. 0.02 8. 26.27 9. 6.44

10. 1.007 11. 0.333 12. 9.000 13. 0.1235 14. 0.0001 15. 2.2112 16. 35.2

17. 6.033 18. 0.04 19. 5 20. 210 21. 0.0045 22. 393 23. 72 24. 60 25. 10

Lesson 4 Homework 3

1. 23 2. 0 3. 8 4. 0.2 5. 6.7 6. 0.1 7. 2.39 8. 0.001 9. 16.71

10. 34.023 11. 0.667 12. 0.001 13. 1.2346 14. 0.0000 15. 22.0988 16. 0.87

17. 400 18. 24.8 19. 0.01 20. 6 21. 39.673 22. 101 23. 21 24. 33 25. 673

Lesson 5 Homework 1

1. 1.058 2. 5.729 3. 32.31 4. 41.62 5. 10.068 6. 0.00213 7. 48.74

8. 3.785 9. 382.656 10. 31.376 11. 49.25 12. 12.554 13. 40.743

14. 23.314 15. 0.198 16. 28.28 17. 15.78 18. 15.54 19. 0.363 20. 177.217

21. 846.053 22. 754.91 23. 26.337 24. 97.868

Lesson 5 Homework 2

1. 1.15 2. 0.029 3. 7.599 4. 6.38 5. 13.007 6. 0.393 7. 112.76

8. 2.596 9. 178.278 10. 23.738 11. 11.08 12. 36.108 13. 38.685 14. 101

15. 2.107 16. 21.003 17. 0.457 18. 7.234 19. 0.018 20. 81.47

21. 26.542 22. 32.4 23. 55.865 24. 8.886

Lesson 6 Homework 1

1. \$41.25 2. \$8.75 3. The sister is 1.26 meters tall, the brother is 1.71 meters tall

4. 0.87 is closer 5. 76.3 feet 6. 21.6 meters 7. \$18 8. \$22.5 9. \$38.10

10. 16.6 minutes

Chapter 7 Test

1. 0.7 2. 0.23 3. 3.037 4. 7.46708 5. forty-five thousandths = 0+4x0.01+5x0.001

6. two and thirty-four thousand eight hundred seventy-five hundred-thousandths =
2+3x0.1+4x0.01+8x0.001+7x0.0001+5x0.00005

7. twelve and seven thousand five hundred thirty-two millionths =
1x10 + 2x1 + 7x0.001 + 5x0.0001 + 3x0.00001 + 2x0.000001

8-11.

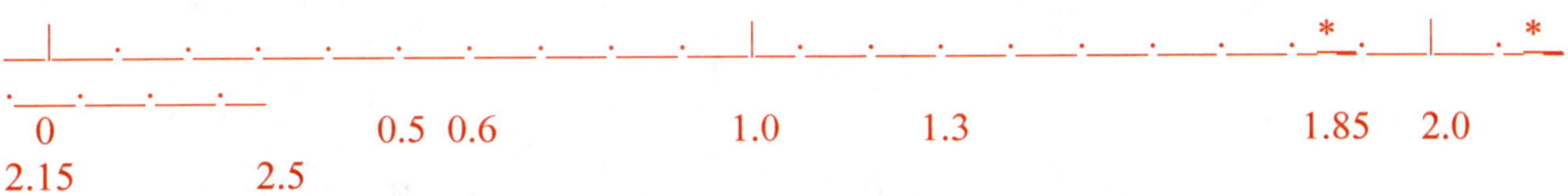

12. < 13. > 14. = 15. 0.9 16. 1.23 17. 73 18. 0.005 19. 3.02

20. 2.718 21. 32.408 22. 52.644 23. 32.917 24. 28.307 25. \$108.81

26. 201.25 lb

Chapter 8

Lesson 1 — Homework 1

1. 3.36 2. 0.072 3. 190.2 4. 7.82 5. 0.927 6. 9.996 7. 60.96 8. 12.728

9. 0.54 10. 0.8 11. 623 12. 1350 13. 867 14. 2.114 15. 0.2028

16. 79.55 17. 0.0048 18. 720.9 19. 5,362.8 20. 48 21. 90 22. 8

23. 3280 24. 250 25. 1.5

Lesson 1 — Homework 2

1. 0.0056 2. 30.26 3. 16.02 4. 0.036 5. 19.688 6. 1.2754 7. 5.7638

8. 0.2592 9. 2.72 10. 324 11. 4.8 12. 0.75 13. 65,780 14. 0.1032

15. 0.0009 16. 0.00632 17. 0.2065 18. 0.000063 19. 0.96 20. 45 21. 2.5

22. 15 23. 140 24. 500 25. 30

Lesson 2 — Homework 1

1. 4.23 2. 1.39 3. 3.03 4. 4.3 5. 0.059 6. 0.652 7. 4.75 8. 0.207

9. 0.0133 10. 3.209 11. 16.08 12. 0.033 13. 0.243 14. 0.109 15. 5.9

16. 2.34 17. 0.0021 18. 0.084 19. 0.0434 20. 6.049 21. 88.9 22. 0.025

23. 0.24 24. 29.78 25. 0.157 26. 0.68 27. 0.124

Lesson 2 — Homework 2

1. 0.446 2. 0.055 3. 16.79 4. 39.96 5. 0.354 6. 1.59 7. 0.0098 8. 0.789

9. 0.073 10. 0.113 11. 0.401 12. 0.036 13. 2.48 14. 0.218 15. 87.1

16. 3.33 17. 0.014 18. 0.0094 19. 0.55 20. 0.341 21. 2.15 22. 0.024

23. 0.105 24. 0.0428 25. 2.21 26. 0.124 27. 0.0069

Lesson 3 Homework 1

1. $0.\overline{7}$ 2. $0.\overline{23}$ 3. $2.7\overline{83}$ 4. $6.\overline{6}$ 5. $56.\overline{247}$ 6. $0.\overline{03}$ 7. $0.2\overline{3}$ 8. $0.58\overline{3}$

9. $0.5\overline{2}$ 10. $0.0\overline{45}$ 11. $5.8\overline{3}$ 12. $12.8\overline{6}$ 13. $8.41\overline{6}$ 14. $0.0\overline{142857}$ 15. $0.2\overline{4}$

16. $0.1\overline{6}$ 17. $1.\overline{81}$ 18. $0.\overline{8}$ 19. 1.625 20. 12.75 21. $3.8\overline{3}$ 22. $0.\overline{857142}$

23. $5.91\overline{6}$ 24. 2.8125

Lesson 3 Homework 2

1. $0.0\overline{5}$ 2. $7.\overline{37}$ 3. $22.\overline{345}$ 4. $0.07\overline{6}$ 5. $787.\overline{87}$ 6. $0.\overline{034}$ 7. $0.08\overline{3}$

8. $6.7\overline{6}$ 9. $0.\overline{285714}$ 10. $1.7\overline{4}$ 11. $0.208\overline{3}$ 12. $0.\overline{027}$ 13. $0.00\overline{6}$ 14. $3.\overline{3}$

15. $2.1\overline{6}$ 16. $0.\overline{6}$ 17. $1.1\overline{6}$ 18. $5.\overline{7}$ 19. 0.375 20. $5.\overline{285714}$ 21. $20.\overline{72}$

22. 5.0625 23. $0.9\overline{3}$ 24. 2.5

Lesson 4 Homework 1

1. 2 2. 33 3. 20 4. 1 5. 2 6. 51 7. 0.8 8. 8.3 9. 0.2 10. 3.2

11. 0.0 12. 94.8 13. 0.45 14. 0.03 15. 6.18 16. 7.71 17. 0.39

18. 0.01 19. 0.078 20. 32.333 21. 0.633 22. 3.614 23. 4.315 24. 0.008

25. 0.2233 26. 0.0055 27. 0.4933

Lesson 4 Homework 2

1. 1 2. 9 3. 0 4. 3 5. 32 6. 3 7. 1.3 8. 4.2 9. 0.0 10. 1.9

11. 0.0 12. 15.5 13. 1.17 14. 0.11 15. 2.11 16. 0.02 17. 2.43

18. 0.04 19. 1.429 20. 0.247 21. 0.303 22. 0.031 23. 0.0693 24. 0.833

25. 0.01457 26. 0.4714 27. 2.6375 28. 0.0161 29. 5.5625 30. 0.0291

Lesson 5 Homework 1

1. 4 2. 0.08 3. 30 4. 0.07 5. 140 6. 0.58 7. 2.07 8. 0.25 9. 1500

10. 0.01 11. 0.0038 12. 5,600 13. 3,400 14. 0.2 15. 0.8 16. 305

17. 1 18. 206 19. 0 20. 3 21. 7 22. 10 23. 0.1

Lesson 5 Homework 2

1. 54 2. 230 3. 1.5 4. 0.33 5. 34 6. 0.73 7. 2400 8. 0.22 9. 2.6

10. 0.0002 11. 3,000 12. 0.34 13. 49 14. 41.9 15. 170.3 16. 14.3 17. 39.1

18. 67 19. 75 20. 6,202 21. 6 22. 4 23. 20 24. 130 25. 0.4

Lesson 5 Homework 3

1. 3.5 2. 7.5 3. 21 4. 8.6 5. 0.54 6. 220 7. 0.203 8. 6.5 9. 0.505

10. 0.0125 11. 7.8 12. 6.6 13. 9.4 14. 0.68 15. 504 16. 2.96

Lesson 6 Homework 1

1. \$5.25 2. \$15.82 3. 154.71 miles 4. \$129 5. 0.12 m/min 6. sending cards abroad is more expensive 7. 6.75 hours 8. 1.25 pages 9. 7.75 10. 13

Chapter 8 Test

1. 2.72 2. 1.504 3. 21.141 4. 0.0054 5. 295.3 6. 305.8 7. 0.8436

8. 1.56 9. 0.063 10. 0.335 11. 5.23 12. 0.84 13. 41 14. 25 15. $0.\overline{3}$

16. $0.8\overline{3}$ 17. 4.375 18. $7.91\overline{6}$ 19. 6.7 20. 0.4 21. 4.6 22. \$0.025 23. \$28

Chapter 9

Lesson 1 — Homework 1

1. $\frac{3}{5}$ 2. 3 3. $\frac{3}{8}$ 4. $\frac{7}{24}$ 5. $\frac{7}{10}$ 6. $4\frac{1}{2}$ 7. $\frac{2}{3}$ 8. $10\frac{9}{10}$ 9. $5\frac{1}{2}$ 10. $7\frac{1}{7}$

Lesson 1 — Homework 2

1. 1 2. $2\frac{1}{2}$ 3. $\frac{1}{36}$ 4. 4 5. 8 6. 9 7. 28 8. 1 9. 5 10. 66

Lesson 2 — Homework 1

1. 1.5 2. 4.3 3. 0.7 4. 0.28 5. 0.51 6. 8 7. 5 8. 1 9. 3 10. 0.076

11. 1 12. 500

Lesson 2 — Homework 2

1. 5 2. 25.1 3. 13 4. 0.32 5. 0.76 6. 0.126 7. 37.5 8. 100 9. 0

10. 1 11. 6 12. 1

Lesson 3 — Homework 1

1. 5.7 2. 0.815 3. 13.8 4. 6.8 5. $9\frac{3}{5}$ 6. 17.1 7. $\frac{17}{40}$ 8. 0.52 9. $\frac{2}{3}$

10. 1.5

Lesson 3 — Homework 2

1. 1 2. $3\frac{1}{4}$ 3. 9 4. 5.2 5. 21 6. 20.775 7. 0.1 8. $1\frac{23}{35}$ 9. 2 10. 10,000

Lesson 4 — Homework 1

1. 11; 12; 13 2. 5.5 miles 3. 10 candies 4. $75 5. $2.40 6. 6.75 hours

7. 25; 20; 15 8. 28 9. $\frac{8}{12}$ 10. 29 boys, 21 girls

Chapter 9 Test

1. 9 2. $6\frac{1}{4}$ 3. 1 4. 28 5. 13.4 6. 60 7. 0 8. 20 9. $\frac{7}{30}$ 10. \$66.67

Chapter 10

Lesson 1 Homework 1

1. 8 2. 15 3. 52 4. 41 5. 0 6. 125 7. –23 8. 14 9. 100 10. –44

11. –73 12. –10 13. –6 14. –4 15. –3 16. 3 17. 5 18. –13 19. 12 20. –5

Lesson 2 Homework 1

1. < 2. > 3. < 4. = 5. > 6. < 7. –12, -3, 0, 9, |–16| 8. –38, -7, 7, |–22|, 45

9. |–36|, 12, 2, -8 , -15 10. 70, |– 50|, 40, -20 , -30

11-14

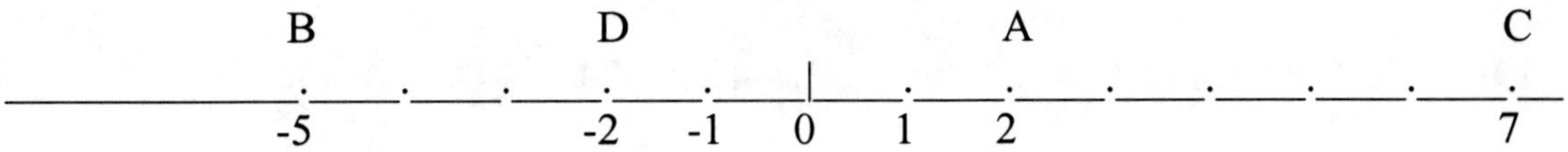

B,D,A,C

15. –6 16. 5 17. –5

Lesson 3 Homework 1

1. –19 2. –5 3. –8 4. 61 5. –59 6. –181 7. –4 8. –23 9. 0

10. 120 11. –55 12. –56 13. 87 14. –132 15. 23 16. 108 17. 142

18. –112 19. –82 20. –82 21 –130 22. 128 23. –77 24. –198 25. 82

26. 0 27. – 178 28. 68 29. 363 30. –59

Lesson 4 Homework 1

1. –4 2. –8 3. 12 4. –15 5. –8 6. 20 7. –31 8. 8 9. 27

10. –35 11. –17 12. –62 13. 63 14. –46 15. 16 16. 0 17. –104

18. –1 19. –25 20. 31 21. 28 22. –121 23. –6 24. 123 25. 112

26. –40 27. –143 28. –45 29. 137 30. 100

ıon 4 Homework 2

1. –16 2. –62 3. –67 4. 34 5. 64 6. 17 7. –61 8. –46 9. 0

10. 16 11. 31 12. –78 13. 62 14. –77 15. –84 16. –94 17. 82

18. 15 19. 19 20. 177 21. 58 22. –100 23. –90 24. –87 25. –35

26. 50 27. 65 28. –94 29. –58 30. 27

Lesson 4 Homework 3

1. –14 2. –8 3. –91 4. –74 5. 68 6. 34 7. 87 8. 36 9. –42

10. –90 11. –29 12. 45 13. –28 14. –83 15. –78 16. –53 17. –75

18. 85 19. 6 20. 46 21. 64 22. 56 23. –160 24. –44 25. 136

26. –120 27. 400 28. –66 29. 4 30. –76

Lesson 5 Homework 1

1. –96 2. –78 3. 180 4. 225 5. -350 6. 144 7. 121 8. -195 9. 448

10. –294 11. 198 12. –152 13. –196 14. 342 15. –676 16. –1,200

17. –120 18. 1,050 19. 1330 20. –864 21. –60 22. 140 23. –720

24. –378 25. –80 26. 252 27. –420 28. –264 29. 24 30. 10,530

Lesson 6 Homework 1

1. –4 2. 7 3. 3 4. –7 5. 7 6. –100 7. 0 8. 3 9. –5 10. 8 11. –15

12. –3 13. 9 14. –5 15. –17 16. 61 17. 8 18. –9 19. –11 20. 13

21. 2 22. 5 23. 3 24. –8 25. –2 26. –4 27. 5 28. 15 29. –15 30. –5

Lesson 7 Homework 1

1. 18 2. –5 3. 7 4. –140 5. 16 6. –26 7. –16 8. –36 9. –8 10. 27

11. –62 12. –8 13. 25 14. –8 15. 18 16. –39

Lesson 7 Homework 2

1. 80 2. –25 3. –4 4. 11 5. –21 6. –5 7. –13 8. 6 9. 80 10. 0

11. 7 12. –9 13. 5 14. –8 15. –1 16. 93

Lesson 8 Homework 1

1. –20 dollars 2. 79 points 3. –1 4. –8, -11, -14 5. 28 6. –7 7. –12, -3

8. 19 9. 5 hours 10. 56

Chapter 10 Test

1. |–16|, 7, 4, -1, -4 2. |–44|, 22, 2, -11, -33 3. 6 4. –43 5. –29

6. –62 7. 75 8. –34 9. –60 10. 12 11. –182 12. 324 13. -4 14. –21

15. 60 16. 39 17. –13 18. –32 19. 11:00 AM 20. 12 21. -2

ORDER FORM

Please send me:

Title	Author	Price	Quantity
Mathematics for Little Ones	Dina Migachyov	$19.95	
Fractions and More	Dina Migachyov	$24.99	

I understand that I may return the books for a full refund – for any reason, no questions asked.

Shipping: $4.00 for the first book and $2.00 for each additional book.

Sales Tax: Please add 7.75% ($1.55/ per book for *Mathematics for Little Ones*, $1.94/per book for *Fractions and More*) for books shipped to Texas addresses.

Please make check payable to: Quaternion Press

Address: Quaternion Press
25018 Granite Path
San Antonio, Texas 78258

Phone: 210-497-1096